CHASIN

Chasing Dragons betwee

Chasing Dragons between Dimensions
An Exploration of Fractals: Mathematics, Philosophy, and Reality

by Matthew Emmanuel Weinberg

© 2023 by Matthew Emmanuel Weinberg

All rights reserved. This book or any portion thereof may not be reproduced or used in any manner whatsoever without the express written permission of the publisher except for the use of brief quotations in a book review.

ISBN: 9798851216251

Preface

Socrates: Now if it is true that the living come from the dead, then our souls must exist in the other world, for if not, how could they have been born again? [...] The soul, then, as being immortal, and having been born again many times, and having seen all things that exist, whether in this world or in the world below, has knowledge of them all; and it is no wonder that he should be able to call to remembrance all that he ever knew about virtue, and about everything else; for as all nature is akin, and the soul has learned all things; there is no difficulty in him eliciting or as men say learning, out of a single recollection all the rest, if a man is strenuous and does not faint; for all inquiry and all learning is but recollection.

Cebes: What you are saying, Socrates, seems to be very much like the truth.

—*Phaedo*

It was a chilly Tuesday afternoon, as I recall, when I stumbled upon an old geometry textbook buried deep in the cavernous corners of my attic. Dust particles danced in the narrow shafts of sunlight that pierced the cobweb-laden window. A strange sense of nostalgia washed over me as I brushed off the layers of time from the book's cover. This was more than a textbook; it was a time machine that transported me back to my own days as a student, when mathematics was a magical world waiting to be explored. Little did I know that this chance encounter would rekindle an old fascination and set me off on an extraordinary journey.

As a math teacher, I had spent countless hours elucidating the elegant dance of numbers and equations. I had taken young minds on tours through the mesmerizing landscape of algebra, guided them through the enigmatic alleys of calculus, and watched their eyes light up as they unraveled the mysteries of trigonometry. But there, in the quiet of my attic, I rediscovered a chapter from the realm of geometry that was unlike anything else: the remarkable world of fractals.

A fractal, as I was reminded that afternoon, is not just an abstract mathematical concept or a geometric shape. It is a window into a realm where intuition and imagination take the driver's seat and conventional rules of dimensions take the back seat. The textbook defined fractals as shapes that are "self-similar"—patterns that repeat themselves at increasingly smaller scales, ad infinitum. But to me they were magic. They were the rabbits pulled out of a mathematician's hat, the portals to a world that was as mind-bending as it was beautiful.

As I flipped through the yellowing pages, memories flooded back. I remembered the first time I saw the spiraling whirls of the Mandelbrot set, a famous example of a fractal. Its endless intricacy, born from a simple mathematical equation, was as mesmerizing as a cosmic dance of celestial bodies. It was a marvel of mathematics that could give one the impression of soaring through an interdimensional vortex. And yet it was not confined to the abstract world of numbers and equations. Fractals, I knew, were woven into the very fabric of our natural world—in the spirals of galaxies, in the branching of trees, in the formation of clouds, and even in the rhythm of our hearts.

Revisiting fractals was like reuniting with an old friend. But it also felt like meeting a celebrity face to face after having seen them only on a screen. I had taught about fractals in my classes, of course, but always as a fleeting, somewhat esoteric part of the curriculum—too complex for a deep dive, too strange for conventional pedagogy. But that afternoon, as the sunlight began to wane and the shadows in the attic grew longer, I felt an irresistible urge to understand them more deeply, to explore their mysteries, and to share their magic with others.

And so my interdimensional odyssey began—an odyssey that was not just about understanding a mathematical concept, but about weaving a tapestry of connections across disciplines, about finding echoes of fractals in art, nature, technology, and even philosophy. It was about seeing the world through a new lens—a fractal lens—and discovering a sense of wonder and curiosity that I thought I had left behind in my own school days.

This book is the result of that odyssey. It is a love letter to the enigmatic world of fractals, an invitation to you, dear reader, to join me on this rollercoaster ride through infinite dimensions. It is an attempt to make the complex simple, the abstract tangible, and the infinite accessible. It's a journey that will take us from the spiraling patterns of Romanesco broccoli to the swirling artistry of Vincent van Gogh, from the rhythmic fluctuations of the stock market to the intricate beauty of a snowflake.

Along the way, we'll delve into the history of fractals, meet the brilliant minds that discovered and explored them, and learn about their applications in various fields. We'll unravel the philosophical implications of their infinite complexity and discuss the ethical questions they raise. We'll even try to comprehend the tantalizing concept of fractional dimensions and the fascinating interplay between order and chaos that fractals embody.

But most importantly, we'll discover the joy of looking at the world through a fractal lens, of finding patterns in the seemingly random, and of seeing unity in diversity. We'll learn to appreciate the magic of mathematics, not as a dry, rigid discipline, but as a language of nature, a tool for understanding the universe, and a source of endless wonder and delight.

My hope is that this book will inspire you, as it did for me, to see fractals not just as mathematical curiosities, but as a testament to the beauty, complexity, and interconnectedness of our world. I hope it sparks in you a sense of curiosity and a thirst for knowledge, and that it makes you see the magic that's hidden in plain sight in the world around us and within us.

So buckle up for an interdimensional rollercoaster ride. Get ready to warp your mind, expand your horizons, and dive headfirst into the captivating world of fractals. The journey is bound to be as fascinating as the destination, and I am thrilled to be your guide.

Let the fractal fantasia begin!

Introduction

Clouds are not spheres, mountains are not cones, coastlines are not circles, and bark is not smooth, nor does lightning travel in a straight line...Nature exhibits not simply a higher degree but an altogether different level of complexity. The number of distinct scales of length of natural patterns is for all practical purposes infinite...The existence of these patterns challenges us to study those forms that Euclid leaves aside as being formless, to investigate the morphology of the amorphous. Mathematicians have disdained this challenge, however, and have increasingly chosen to flee from nature by devising theories unrelated to anything we can see or feel.
—Benoît Mandelbrot

Welcome to the world of fractals!

Prepare yourself for an enchanting journey, a labyrinthine dance through the fine threads of reality: a place where mathematics, nature, and imagination converge. Welcome to the mesmerizing world of fractals, an interdimensional landscape where patterns, complexity, and beauty coalesce into something that will surely astound you.

The term "fractal," coined by mathematician Benoît Mandelbrot in 1975, might sound alien, yet it represents a concept that permeates our reality. From the intricate branching of trees and the patterns in a shell to the mysterious pathways of lightning and even the structure of galaxies, fractals are ubiquitous, offering a lens through which to view the world's complex symmetry. It's a unique perspective that unveils the underlying order hidden within the apparent chaos of nature and the universe.

In essence, a fractal is a geometric figure, but it's not your everyday geometry. Unlike the perfect circles, squares, and triangles of Euclidean geometry, fractals embody a world of shapes that are irregular, fragmented, and infinitely complex. Yet amid this complexity lies a profound simplicity: a self-similarity where each part, no matter how small, mirrors the whole. Zoom in or zoom out, and the pattern repeats, unfolding like a never-ending Russian doll.

This self-similarity, this repetition across scales, is one of the most mind-boggling features of fractals. It's a paradox where the finite harbors the infinite, where simplicity breeds complexity. This infinite complexity within finite bounds is also known as the fractal dimension, a dimension that exists "in between" the standard whole-number dimensions we're familiar with, creating a magical realm of fractional dimensions.

Fractals, however, are not just about pretty shapes and patterns. The ideas they encapsulate offer profound insights into the very nature of our reality. Fractals challenge our traditional concepts of space and time, nudging us to embrace the fluid, dynamic, and interconnected nature of existence. They ask us to consider a universe not of isolated objects and linear progressions, but of interconnected patterns and recursive processes, where everything affects everything else in a delicate dance of cause and effect.

These ideas align with some of the most profound philosophical musings throughout history. In the echoes of Plato's dialogues, we find the idea that "all learning is but recollection," resonating with the fractal notion of self-similarity. And in the teachings of Eastern philosophies, we find parallels with the fractal's interwoven dance of simplicity and complexity, chaos and order.

Even number theory, a branch of mathematics known for its abstraction, finds a companion in fractals. Fractals breathe life into numbers, adding dimensions and complexity to the otherwise flat landscape of pure mathematics. They reveal the music hidden within the dry notes of numbers, painting a vivid picture of the symphony that numbers can create.

The history of fractals is as fascinating as the concept itself. It's a story of visionaries and mavericks, of thinkers who dared to venture beyond the traditional boundaries of thought.

From the seventeenth-century philosopher and mathematician Gottfried Leibniz, who pondered on the idea of self-similarity, to the nineteenth-century mathematicians like Karl Weierstrass, who introduced functions that were "everywhere continuous but nowhere differentiable," and finally to Benoît Mandelbrot, who gave the concept its name and its rightful place in mathematics.

The journey of fractals is a testament to human curiosity and ingenuity. And the journey doesn't end there. Fractals have found their way into various applications, proving their worth not just as a theoretical concept but as a practical tool. From computer graphics and image compression to the study of markets in financial analysis, from understanding the unpredictable patterns of weather to the modeling of biological processes and population patterns, fractals are becoming an indispensable tool in our quest to understand and navigate the world.

The most compelling part of this journey into the world of fractals is how it weaves together threads from diverse domains, creating a tapestry that reflects the interconnectedness of our universe. The beauty of a fractal lies not just in its visual appeal, but in its ability to connect, to resonate, to echo the fundamental structures and processes of reality.

This is not a journey for mathematicians alone. It's a journey for anyone who appreciates beauty, who is fascinated by the complexity of nature, who marvels at the intricacies of existence. It's a journey for the curious, the thinkers, the dreamers. It's a journey that transcends disciplines, cultures, and epochs, linking us with the timeless quest of humanity to understand the cosmos.

Yet as complex and profound as the world of fractals may be, it is remarkably accessible. You don't need to be a mathematician to appreciate the beauty of fractals or grasp the insights they offer. All you need is curiosity, an open mind, and a willingness to see the world from a different perspective.

As we delve deeper into the world of fractals in

the following chapters, we'll explore their origins, their properties, their philosophical implications, and their practical applications. We'll see how they challenge our conventional wisdom, reshape our understanding, and offer a new way of seeing the world.

So are you ready to embark on this interdimensional rollercoaster ride? Are you ready to delve into the endless, intricate, and fascinating world of fractals? Let's begin this journey together, into the depths of the world within and beyond, into the heart of the fractal universe.

Welcome to the world of fractals, a place where mathematics meets philosophy, where art meets science, where the finite and the infinite dance in a ballet of cosmic harmony. Welcome to a world where the complexity of existence is mirrored in the simplicity of shapes, where the vastness of the cosmos is echoed in the smallest fragment. Welcome to a world that, once entered, will forever change how you perceive the universe around you.

Why Should I Care About Fractals?

As we peer into the looking glass of the world of fractals, you might be asking, "Why should I care about fractals?" After all, they may seem like a mathematical curiosity, a fanciful detour from the practical concerns of day-to-day life. But fractals, in their intricate beauty and complexity, have far-reaching implications and applications that go beyond the realm of abstract mathematics. They offer a new way of understanding the world, a perspective that has the potential to change how we approach everything from science and technology to art and philosophy.

Let's start with the realm of science and technology. In this domain, fractals are much more than a theoretical concept —they are a practical tool. Fractals have found applications in diverse areas, illuminating the path to novel solutions and innovative approaches.

Take, for instance, the field of computer graphics. Fractals have been used to create stunningly realistic virtual landscapes in films and video games. These virtual environments, with their intricate details and endless variations, would be nearly impossible to create without the self-similar properties of fractals. By using fractal algorithms, graphic designers can generate complex shapes and forms with a minimal set of instructions, saving time and computational resources.

In the realm of telecommunications, fractals have revolutionized antenna design. Fractal-shaped antennas can operate at multiple frequencies, making them highly efficient and versatile. This has significant implications for mobile communication devices, improving their performance and reducing their size.

Beyond these practical applications, fractals have a profound impact on our understanding of nature and the universe. The patterns and processes we observe in the world around us—from the branching of trees to the formation of clouds, from the distribution of galaxies to the rhythm of heartbeats—often exhibit fractal characteristics. By studying these fractal patterns, we can gain insights into the underlying principles that govern these phenomena. Fractals, therefore, offer us a mathematical language to describe the world, a language that captures the complexity and beauty of nature in a way that traditional mathematical models cannot.

But the influence of fractals extends even further. They have found their way into the realm of art and design, offering artists a new palette of shapes and forms to explore. The mesmerizing patterns and infinite complexity of fractals have inspired artists, leading to a whole new genre of fractal art. This synergy between mathematics and art exemplifies the aesthetic appeal of fractals, revealing the inherent beauty in the laws of nature.

On a deeper level, fractals challenge our conventional notions of space, time, and reality. They compel us to consider a universe that is composed not of isolated objects but of interconnected patterns. They invite us to embrace a dynamic, fluid, and recursive view of existence, where everything affects everything else in a complex web of interactions. This perspective resonates with some of the most profound philosophical ideas, bridging the gap between science and philosophy, between the tangible and the abstract.

In this sense, fractals are not just about numbers and shapes. They are about the very fabric of our reality. They reflect the complex, interconnected, and dynamic nature of the universe. They embody the dance of chaos and order, the interplay of simplicity and complexity, the balance of finitude and infinity.

So why should you care about fractals? Because they offer a fresh, innovative, and profound perspective on the world. Because they bridge the gap between diverse fields, bringing together science, technology, art, and philosophy in a harmonious dance. Because they reveal the beauty, complexity, and interconnectedness of the universe.

In the following chapters, we will delve deeper into the fascinating world of fractals, exploring their properties, their implications, and their applications. We will see how they reshape our understanding of the world and how they inspire innovation in various fields. Whether you're a scientist, a philosopher, an artist, or simply a curious mind, fractals have something to offer you. They invite you to step beyond the familiar, to embrace the complexity, to revel in the beauty of the universe.

Fractals also invite us to reflect on the nature of knowledge itself. They remind us that understanding is not about simplifying the complex into the simple, but about appreciating the complex in its own right. They encourage us to move away from linear thinking and embrace a more holistic, interconnected perspective. This shift in thinking has the potential to change not only how we understand the world, but also how we interact with it.

Moreover, fractals are a testament to the power of curiosity and imagination. They emerged from the efforts of thinkers who dared to venture beyond the traditional boundaries of thought, who dared to imagine a different kind of geometry, a different kind of reality. Their story is a reminder that progress often comes from challenging the status quo, from embracing the unconventional, from daring to dream.

So why should you care about fractals? Because they are more than just a mathematical concept. They are a lens through which to view the world, a tool to understand it, a language to describe it, and an inspiration to engage with it. They reflect the beauty, complexity, and interconnectedness of the universe. They offer a fresh, innovative, and profound perspective on reality. They encourage us to be curious, to be imaginative, to be daring.

As we continue our journey into the world of fractals, remember that this is not just a journey into the realm of mathematics. It's a journey into the depths of reality, into the heart of existence. It's a journey that will take us beyond the familiar, beyond the conventional, into a world that is complex, beautiful, and profoundly interconnected.

In the end, caring about fractals is about more than understanding a mathematical concept. It's about embracing a new perspective on the world, a perspective that recognizes the beauty in complexity, the order in chaos, the unity in diversity. It's about appreciating the intricate dance of the universe, the delicate balance of simplicity and complexity, the endless interplay of the finite and the infinite.

So buckle up and get ready for an exciting exploration into the world of fractals, a journey that will illuminate your mind, stir your imagination, and ignite your curiosity. Welcome to the world of fractals, a world that is sure to leave you amazed, inspired, and profoundly changed.

How to Read This Book

As you prepare to embark on this fascinating journey into the world of fractals, it's important to set the stage for how best to approach the reading of this book. *Chasing Dragons between*

Dimensions is designed to be more than just a dry academic text or a simple reference book. It's intended to be a dynamic, interactive exploration of a rich and complex field.

Here's some guidance on how to get the most out of your journey:

1. **Start with an Open Mind:** This book is not just about mathematics; it's about a new way of perceiving the world. It's about understanding the inherent patterns and structures that permeate our universe. So begin your journey with an open mind. Be prepared to encounter new concepts and to challenge your preconceptions.
2. **Embrace the Complexity:** Fractals are complex, but so is the world we live in. Don't be intimidated by the complexity; instead, embrace it. Remember, the aim is not to oversimplify, but to appreciate the beauty and intricacy of complexity.
3. **Engage Actively:** This book is structured to be interactive. Each chapter builds on the previous one, and many concepts are revisited and expanded upon as we delve deeper into the world of fractals. Take notes, ponder the ideas presented, and don't hesitate to revisit earlier sections as needed.
4. **Appreciate the Interdisciplinary Approach:** Fractals are inherently interdisciplinary, with connections to mathematics, science, art, philosophy, and more. The book reflects this diversity. Whether you're a mathematician, scientist, artist, philosopher, or simply a curious mind, there's something here for you.
5. **Apply What You Learn:** The book is not just about theory; it also explores practical applications of fractals. As you read, consider how you might apply these concepts in your own life, work, or field of study.
6. **Enjoy the Journey:** Finally, remember that this book is a journey, an adventure into a new realm of understanding. Enjoy the ride! Let yourself be captivated by the beauty of fractals, the wonder of discovery, and the joy

of learning.

This is not a book that should be rushed through. It's a journey to be savored, a treasure trove of insights to be discovered, a universe to be explored. Each chapter, each section, each page is part of the larger voyage into the universe of fractals.

So take your time. Let the ideas simmer in your mind. Reflect on what you've learned. Engage with the material, challenge it, question it. This book is meant to be a conversation, a dialogue between you and the world of fractals.

Now, as we embark on this interdimensional rollercoaster ride together, let's dive into the enchanting, mesmerizing, and intriguing world of fractals. Hold on tight, keep an open mind, and enjoy the journey!

1: UNDERSTANDING BASIC SHAPES AND PATTERNS

She had not gone much farther before she came in sight of the house of the March Hare: she thought it must be the right house, because the chimneys were shaped like ears and the roof was thatched with fur. It was so large a house, that she did not like to go nearer till she had nibbled some more of the left-hand bit of mushroom, and raised herself to about two feet high: even then she walked up towards it rather timidly, saying to herself, "Suppose it should be raving mad after all! I almost wish I'd gone to see the Duchess!"
—Lewis Carroll, *Alice's Adventures in Wonderland*

As we embark on this journey into the world of fractals, it is important to first ground ourselves in the understanding of basicshapes and patterns. After all, it's from these simple forms that the complex, mesmerizing structures of fractals emerge.

To begin with, let's consider what we mean by "shape" and "pattern." A shape, in the simplest terms, is the form of an object or its external boundary. A pattern, on the other hand, is a repeated decorative design, a consistent and recurring characteristic or trait. Shapes and patterns are fundamental to how we perceive and understand the world around us.

Traditionally, our understanding of shapes is rooted in Euclidean geometry—the geometry of flat space that we learn in school. This is the realm of familiar shapes such as circles, squares, and triangles. These shapes have well-defined properties: a circle has a constant radius, a square has equal sides, a triangle has angles that sum to 180 degrees. These shapes are simple, regular, and predictable. They are the building blocks of the Euclidean world.

However, when we look at the world around us, we quickly realize that it is filled with shapes and patterns that are far from simple or regular. Consider the intricate branching of a tree, the jagged outline of a mountain range, the complex interweaving of a river network. These shapes don't conform to the regularity of circles, squares, or triangles. They are complex, irregular, and unpredictable. Yet they too are shapes; they too form patterns.

This is where the concept of fractals comes in. Fractals are mathematical shapes that are self-similar. This means that they look the same at any scale—if you zoom in or out, the shape's character doesn't change. This property of self-similarity gives fractals their characteristic complexity and intricacy.

Fractals are everywhere in nature. The branching of trees is a fractal pattern—each branch is a smaller version of the tree itself. The network of veins in a leaf is a fractal—each vein is a miniaturized copy of the whole. The jagged coastline is a fractal—each inlet and headland is a scaled-down version of the larger shape. These natural shapes are fractals not because they are perfect mathematical forms, but because they exhibit the property of self-similarity.

The concept of fractals extends our understanding of shapes and patterns. It allows us to describe and analyze the complex, irregular forms found in nature. It provides a mathematical framework to understand the seemingly chaotic world around us.

Understanding fractals starts with rethinking our concept of dimension. In Euclidean geometry dimension is a clear-cut concept. A line is one-dimensional, a square is two-dimensional, a cube is three-dimensional. But fractals exist in the spaces between these whole-number dimensions. A fractal curve can have a dimension between one and two; a fractal surface can have a dimension between two and three. This is the concept of fractal dimension, a measure of the complexity of a fractal shape.

This idea of fractal dimension might seem counterintuitive. After all, how can something be "more than a line but less than a surface?" But this is the essence of fractals—they challenge our intuitive notions; they push the boundaries of our understanding.

To comprehend this, consider the Koch snowflake, a classic fractal shape. It starts with an equilateral triangle. Then, on each side of the triangle, we add a smaller equilateral triangle, and so on, infinitely. The resulting shape has a finite area (it can be contained within a circle) but an infinite perimeter (the length of the boundary keeps increasing with each iteration). It exists somewhere between one and two dimensions, neither fully a line nor entirely a plane.

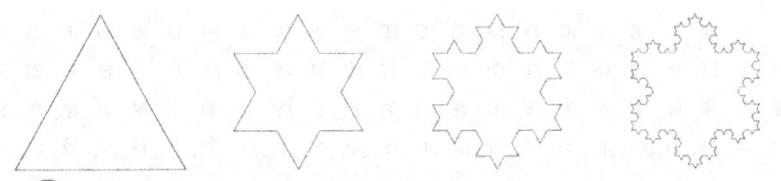

This fascinating property of fractals—to embody more complexity than a line but less than a plane—is central to understanding many of the complex structures in nature and even in human-made systems. Fractals allow us to quantify and analyze this complexity in a precise and meaningful way.

So why are these basic shapes and patterns important? Because they serve as the foundation for the incredible world of fractals. They are the simple beginnings from which the intricate and beautiful structures of fractals emerge.

But fractals are not just about mathematical shapes and patterns. They reflect a fundamental principle of nature: that of self-organization. From the branching of trees to the formation of river networks, nature organizes itself into patterns that are self-similar, that exhibit the property of "scale invariance." Fractals provide a mathematical language to describe this inherent property of nature.

Moreover, understanding fractals requires us to expand our concept of dimension, to move beyond the whole numbers and embrace the spaces in between. This shift in perspective is profound. It challenges our intuitive notions; it pushes the boundaries of our understanding. It reflects the spirit of fractals —to explore the complexity, to embrace the irregularity, to revel in the beauty of the universe.

In the end, the study of fractals is not just about understanding mathematical shapes and patterns. It's about appreciating the beauty and complexity of the world around us. It's about recognizing the inherent patterns and structures that permeate our universe. It's about exploring the depths of reality, about venturing into the unknown, about discovering the infinite in the finite.

As we continue our journey into the world of fractals, let's keep these basic shapes and patterns in mind. Let's remember that from the simple can emerge the complex, that from the predictable can arise the unpredictable, that from the familiar can unfold the unfamiliar. Let's embrace the journey; let's celebrate the exploration; let's revel in the discovery. Welcome to the world of fractals, a world of infinite complexity and endless wonder.

Understanding fractals requires a grasp of certain key ideas in geometry. However, rest assured that we are not about to dive into a dry, technical mathematical treatise. Instead, let's approach these ideas as a traveler in a foreign land might attempt to learn the local language—not to pass a grammar test, but to have richer, deeper interactions with the landscape and its inhabitants.

In our case, the landscape is the enchanting realm of fractals, and its inhabitants are the various shapes and patterns that populate it. To really engage with this land and its people, we need to understand the language of geometry. Let's explore the fundamental principles that will serve as our linguistic guide.

1. **The Concept of Dimension:** Dimension is a fundamental concept in geometry, and it's crucial for understanding fractals. We are familiar with the idea of one, two, and three dimensions from everyday life. A line is one-dimensional; a plane (like a piece of paper) is two-dimensional; and a solid object (like a cube) is three-dimensional. Fractals, however, challenge this simple understanding by existing in the spaces between these whole-number dimensions. This fractal dimension is a measure of the complexity of a fractal shape.

2. **The Idea of Scale:** Scale is another fundamental concept in geometry. In a traditional geometric shape, the scale is fixed; a square remains a square whether we zoom in or out. But fractals exhibit "scale invariance," meaning they look the same at any scale. This property of self-similarity gives fractals their characteristic complexity and beauty.

3. **The Nature of Space:** Euclidean geometry, the geometry we learn in school, is based on flat space. But the space of fractals is not flat; it is "fractal space." This space is filled with intricate shapes and patterns that repeat at different scales, creating a sense of depth and complexity that is unique to fractals.

4. **The Principle of Self-Similarity:** Self-similarity is a defining property of fractals. It means that a shape is made up of smaller copies of itself, each of which is also made up of smaller copies of itself, and so on, potentially ad infinitum. This recursive process generates the intricate and beautiful patterns that we associate with fractals.

5. **The Notion of Infinity:** Infinity is a concept that is central to fractals. Fractal shapes are generated by repeating a simple process an infinite number of times. This gives fractals their characteristic property of infinite complexity

—the deeper we delve into a fractal, the more detail we discover.

Together, these concepts form the language of fractal geometry. They are the keys to unlocking the beauty and complexity of the fractal landscape. But they are more than just mathematical ideas; they are philosophical principles that reflect the inherent patterns and structures of the universe.

Understanding these principles allows us to engage more deeply with the world of fractals. It enables us to appreciate the intricate shapes and patterns that populate this world, to marvel at their beauty and complexity, to delve into their infinite depths.

So as we continue our journey into the realm of fractals, let's keep these principles in mind. Let's remember that the language of geometry is our guide, that these fundamental concepts are our map. Let's embrace the exploration; let's celebrate the discovery; let's revel in the journey. Welcome to the world of fractals, a world of endless wonder and infinite beauty.

As we journey further into the universe of fractals, our language of geometry requires a few additional words and phrases. Like adding spices to a dish to enhance its flavor, these advanced concepts add depth and nuance to our understanding of fractals. They can seem a bit intimidating at first, just like unfamiliar spices, but they become less daunting and more intriguing as we begin to see their effects in the fractal world.

1. **Iteration:** Iteration is the process of repeating a certain set of mathematical operations over and over. This may seem like a simple concept, but it's at the heart of fractal generation. With each iteration, the fractal shape becomes more complex, more intricate. It's like folding a piece of paper: each fold adds a new layer of complexity. Through the power of iteration, a simple shape can transform into a fractal of astonishing beauty and detail.

2. **Infinity and Limits:** In a fractal, the process of iteration goes on forever; it is infinite. This is where the concept of limits becomes crucial. A limit, in mathematical

terms, is the value that a sequence approaches as it goes on indefinitely. The limit concept helps us make sense of the infinite iterations in fractals. It's like trying to reach a destination by continually halving the distance: you never quite get there, but you get closer and closer.

3. **Non-Euclidean Geometry:** Euclidean geometry is the world of flat space, straight lines, and perfect circles. Fractals, however, dwell in the realm of non-Euclidean geometry. This is a geometry of curved space, where the shortest distance between two points is not necessarily a straight line. In non-Euclidean geometry, parallel lines can meet, and the angles of a triangle don't always add up to 180 degrees. It's a bit like Alice's Wonderland, where the ordinary rules of geometry are turned on their head.

4. **Complex Numbers:** Complex numbers are numbers that involve the square root of negative one, represented as i. While this might seem strange and unreal, complex numbers play a crucial role in generating certain types of fractals, like the Mandelbrot set. They are like the imaginary friends of numbers—you can't see them, but they can create wonderful things.

5. **Chaos and Order:** Fractals embody a delicate balance between chaos and order. On the one hand, they are generated by simple, deterministic rules—that's the order. On the other hand, the outcome of these rules is highly sensitive to the initial conditions—that's the chaos. This interplay between chaos and order is a hallmark of fractals, and it's what gives them their unique beauty and complexity.

Together, these advanced concepts provide us with a richer, deeper language to explore the fractal universe. They allow us to appreciate the intricate patterns, the infinite complexity, and the delicate balance between chaos and order that define fractals.

So as we venture further into the fractal landscape, let's remember these advanced concepts. Let's use them as our guide, as our map. And most importantly, let's keep our sense of wonder and curiosity alive. For the fractal universe is a place of endless discovery, of infinite beauty, of eternal wonder. And we are just beginning our journey.

Meet the Fractals: Infinite Complexity in Simple Shapes
Welcome back to our exploration of the fractal universe! We've equipped ourselves with the necessary vocabulary and ideas from the realm of geometry, both basic and advanced. Now it's time to meet the stars of this show, the fractals themselves. Let's embark on a grand tour of some of the most famous and fascinating fractals and see how they embody the geometric principles we've discussed.

Our first stop is the **Sierpiński triangle**. It starts as a simple equilateral triangle, but with each iteration, the middle of each smaller triangle is cut out, creating a new set of even smaller triangles. This process goes on indefinitely, resulting in a fractal that is intricate and beautiful. The Sierpiński triangle embodies the principle of self-similarity, with each smaller triangle a reduced-scale copy of the whole. It also demonstrates the concept of a fractal dimension; although it starts as a two-dimensional triangle, its fractal dimension is somewhere between one and two due to its increasing complexity.

Next, we visit the **Koch snowflake**. It begins with an equilateral triangle, and with each iteration, smaller triangles are added to each side, increasing the complexity and length of the boundary. Despite this, the area of the snowflake remains finite, giving us a shape with a finite area but an infinite perimeter—an astonishing paradox! The Koch snowflake exemplifies the idea of infinite complexity derived from simple iterative processes.

Our journey takes us then to the mesmerizing **Mandelbrot set**. Named after the mathematician Benoît Mandelbrot, who did much to popularize fractals, this set is generated using complex numbers and iterative processes. The resulting shape is captivating, with an infinitely complex boundary that reveals ever-more intricate detail the more you zoom in. The Mandelbrot set not only demonstrates self-similarity and infinite complexity, but also brings to life the abstract concepts of complex numbers and limits.

The **Julia set** is our next stop. Like the Mandelbrot set, it's generated using complex numbers, but with a different iterative process. Each Julia set is associated with a particular complex number, and the array of possible Julia sets is as diverse and fascinating as the inhabitants of a coral reef. The Julia sets show us the delicate balance between order (the deterministic iterative process) and chaos (the highly sensitive dependence on the initial complex number).

Finally, we reach the **Cantor set**. It begins as a simple line segment, but with each iteration, the middle third of each segment is removed. What remains is a set of points that is uncountably infinite but has a total length of zero—another mind-boggling paradox! The Cantor set embodies the principles of self-similarity, infinite complexity, and fractal dimension.

These fractals, and many others like them, are the inhabitants of the fractal universe. Each of them tells a story of infinite complexity arising from simple shapes and iterative processes. Each embodies the principles of fractal geometry—self-similarity, scale invariance, fractal dimension, and the interplay between chaos and order.

Meeting these fractals, seeing these principles come to life, gives us a deeper understanding of the fractal universe. It allows us to appreciate the beauty and complexity of this universe, to marvel at its infinite detail, to revel in its paradoxes and mysteries. So, as we continue our journey into the world of fractals, let's remember these inhabitants. Let's recall their stories, their lessons, their beauty. For the fractal universe is a place of endless exploration, a place where the simple and the complex, the finite and the infinite, the ordinary and the extraordinary, intersect and intertwine in the most remarkable ways.

As we traverse the fractal landscape, we will come across the **Hilbert curve**, a fractal that fills an entire two-dimensional space, yet never crosses its own path. This strange creature defies our intuitive understanding of dimensions, showing us that even a one-dimensional curve can completely fill a two-dimensional area.

We'll encounter the **Barnsley fern**, a fractal that mimics the natural patterns of fern leaves. This fractal, generated using a type of fractal known as an iterated function system, reminds us of the deep connections between fractal geometry and the natural world.

We'll gaze upon the **dragon curve**, a fractal generated from a simple process of folding a strip of paper. This fractal reveals the surprising complexity that can emerge from a simple, iterative process. It shows us that, in the fractal universe, the act of folding can be a path to infinite complexity.

And there are many more fractals waiting to be discovered: fractals that haven't been named yet, fractals that haven't been imagined yet. This is the beauty of the fractal universe. It is a universe of endless discovery, a universe where the simple gives birth to the complex, where the known leads to the unknown, where the finite reveals the infinite.

So, as we delve deeper into this universe, let's keep our eyes open, our minds curious, and our hearts full of wonder. Let's appreciate the beauty and complexity of each fractal we encounter, marvel at the infinite detail they reveal, and revel in the mysteries and paradoxes they embody.

Let's remember that each fractal is a story, a story of infinite complexity arising from simple shapes and iterative processes. And let's remember that we, too, are part of this story. For we are not just observers of the fractal universe—we are explorers; we are discoverers; we are storytellers.

In the next part of our journey, we will delve deeper into the mathematical principles that underpin fractals, exploring concepts like self-similarity, scale invariance, and fractal dimensions in greater detail. But for now, let's take a moment to reflect on the fractals we've met, to marvel at their beauty and complexity, and to appreciate the remarkable universe they inhabit.

Stay curious, stay awed, and remember: the world of fractals is as boundless as our capacity to wonder.

Spotlight on Sierpiński Shapes: Triangles and Carpets
Welcome back, fellow fractal explorer! Having familiarized ourselves with a broad cast of fractals, it's time to get up close and personal with a few of them. Our first stop is at the doorstep of a celebrated resident of the fractal universe, the Sierpiński triangle.

The Sierpiński triangle begins its life as a simple equilateral triangle. It could've been content with its three-sided symmetry, but that wouldn't make it a fractal. It yearns for more —more complexity, more detail, more depth. And so it invites the principle of iteration to work its magic.

With each iteration, the Sierpiński triangle does something astonishing: it cuts out the middle of each smaller triangle, creating a new set of even smaller triangles. What was once a solid shape now becomes a lattice of smaller triangles, each a perfect, scaled-down copy of the original. And this process goes on indefinitely, adding layer upon layer of complexity.

What makes the Sierpiński triangle remarkable is not just its intricate beauty, but the geometric principles it embodies. It's a poster child for self-similarity, each smaller triangle being a reduced-scale copy of the whole. And while it starts as a two-dimensional figure, its fractal dimension is somewhere between one and two, thanks to the added complexity of each iteration. This triangle, as it turns out, lives in the fractional dimensions, a place that wasn't on our geometric map until fractals came into the picture.

But the Sierpiński family doesn't stop at triangles. Meet the **Sierpiński carpet**, a fractal that applies the same iterative process not to a triangle, but to a square. It starts as a solid square, but with each iteration, it removes the middle third of each smaller square, creating an intricate pattern of smaller squares.

Like its triangular cousin, the Sierpiński carpet is a testament to the power of iteration and self-similarity. It turns a simple, two-dimensional square into a complex, fractal shape with a dimension somewhere between two and three. And it does so with a simple, repetitive process that anyone could follow.

The Sierpiński triangle and carpet are more than just shapes. They're stories, narratives about how simple shapes can evolve into complex fractals through the power of iteration and self-similarity. They're lessons about the existence of fractional dimensions, a concept that challenges our conventional understanding of geometry.

As we continue our journey into the fractal universe, let's carry these stories and lessons with us. Let's remember the Sierpiński shapes not just for their beauty, but for the principles they embody, the boundaries they push, and the mind-boggling possibilities they introduce.

Next, we'll delve into the frosty world of the Koch snowflake, a fractal that brings a chilly twist to the concept of infinity. But for now, let's take a moment to appreciate the Sierpiński shapes, their elegance, their complexity, and their uncanny ability to turn the simple into the extraordinary.

Journey into the Frost: Koch Snowflake and Other Fractals
Our next stop on this fractal journey takes us to a frosty landscape, a world of ice and snow where the simple and the complex coexist. Here we meet one of the most iconic fractals: the Koch snowflake.

The Koch snowflake begins its life as a humble equilateral triangle. However, like all fractals, it dreams of something more —more complexity, more detail, more depth. So it calls upon the power of iteration to transform itself.

In the first iteration, the Koch snowflake adds a smaller equilateral triangle to each side. With each subsequent iteration, it adds smaller triangles to the new sides, growing more intricate with every step. This process continues indefinitely, with each step unfolding more detail, like a snowflake crystallizing in slow motion.

But the real magic of the Koch snowflake lies in the paradox it embodies. As it grows, the perimeter of the snowflake increases with each iteration, heading toward infinity. Yet despite this, the area of the snowflake, the space it occupies on the plane, remains finite. Here we have a shape that contains a finite area but is bounded by an infinite perimeter—an astonishing paradox indeed!

The Koch snowflake isn't the only fractal in the Koch family. There's also the **Koch Curve**, which starts as a straight line and adds smaller segments with each iteration, creating a jagged, infinitely complex curve. Like the snowflake, the Koch Curve's length heads toward infinity as it iterates, demonstrating the same mind-boggling paradox of infinite length but finite area.

Then there's the **Koch anti-snowflake**, a fascinating variation where the added triangles point inward rather than outward. This creates a fractal with a finite perimeter but an infinite area—a paradox that's the exact opposite of the Koch snowflake!

The Koch fractals, with their intricate beauty and paradoxical properties, are a testament to the magical world of fractals. They show us how a simple iterative process can generate infinite complexity, how a finite shape can have an infinite perimeter or area, and how the simple and the complex can coexist in the same figure.

As we continue our journey, let's carry the lessons of the Koch fractals with us. They remind us that in the world of fractals, the extraordinary can arise from the ordinary, the infinite can emerge from the finite, and the paradoxical can become the norm.

Our next stop will be the realm of the Mandelbrot and Julia sets, where we'll see how complex numbers can generate stunningly beautiful fractals. But for now, let's reflect on the icy world of the Koch fractals, their startling paradoxes, and their intricate beauty.

Navigating the Mandelbrot Universe: Where
Number Theory Meets Fractal Art

We're about to embark on a journey into one of the most famous and visually stunning regions of the fractal universe: the realm of the Mandelbrot set. Here the abstract world of number theory collides with the tangible beauty of fractal art, creating a landscape of infinite complexity and breathtaking beauty.

The Mandelbrot set is not just a fractal; it's a map, a guide to the complex dynamics of quadratic polynomials. It's a fractal explorer's dream, with its infinite coastline filled with bulbs, tendrils, and miniature replicas of the whole set. Named after the mathematician Benoît Mandelbrot, who did extensive work in fractal geometry, the Mandelbrot set is one of the most recognizable fractals.

Unlike the Koch snowflake or the Sierpiński triangle, the Mandelbrot set doesn't start with a simple geometric shape. Instead, it begins with a mathematical function involving complex numbers. A point in the complex plane is part of the Mandelbrot set if, when you apply the function repeatedly to the point, the result doesn't go off to infinity. It's this simple rule, applied over and over again, that generates the incredible complexity of the Mandelbrot set.

The beauty of the Mandelbrot set lies not just in its intricate shape, but also in its deep connections with number theory, dynamical systems, and chaos theory. It's a showcase of the surprising patterns and structures that can emerge from simple mathematical rules.

Associated with the Mandelbrot set are the Julia sets. These are fractals that are generated using a rule similar to the one for the Mandelbrot set, but with a slight twist. For the Julia sets, the rule is applied not just to one point, but to every point in the complex plane. The result is a family of fractals, each with its own unique, mesmerizing pattern.

The Mandelbrot and Julia sets are a powerful demonstration of the infinite complexity and beauty that can emerge from simple mathematical rules. They're a reminder that in the fractal universe, the simple can give rise to the complex, the abstract can manifest as the tangible, and the finite can reveal the infinite.

As we continue our journey, let's remember the Mandelbrot and Julia sets, not just for their beauty, but for the deep mathematical principles they embody. And let's prepare ourselves for our next stop: the world of three-dimensional fractals, where we'll encounter shapes that defy our intuitive understanding of space and dimension. But for now, let's take a moment to appreciate the Mandelbrot and Julia sets, their intricate patterns, and their profound connections to the abstract world of mathematics. In the realm of fractals, these sets stand as a testament to the power of simple rules to generate infinite complexity.

A Visit to Cantor's Paradise: Discovering Infinity in the Cantor Set
Now we venture into a different yet equally fascinating corner of the fractal universe: the realm of the Cantor set. Named after the mathematician Georg Cantor, this fractal is a marvel of simplicity and paradox, a testament to the mind-boggling nature of infinity.

The Cantor set starts with something as ordinary as a line segment. But it won't remain ordinary for long. Like other fractals, the Cantor set seeks the extraordinary in the mundane. Through a process of iterative removal, it transforms itself from a solid line segment into a scatter of dustlike points.

Here's how it works. You start with a line segment, which represents the first stage. In the next stage, you remove the middle third of that segment, leaving two smaller segments. In the third stage, you remove the middle third of each of these smaller segments, and so on. With each stage, more and more of the line is removed, until, in the limit, you're left with a set of points that are disconnected, yet infinitely many.

But the real wonder of the Cantor set lies in the paradoxes it embodies. Despite the fact that you've removed an infinite amount of line, there are still points left over—in fact, an uncountably infinite number of them! The Cantor set teaches us that infinity can come in different sizes, a mind-bending concept that has profound implications in number theory and other areas of mathematics.

Additionally, the Cantor set, despite being made up of disconnected points, is not just a one-dimensional figure. It has a fractal dimension between zero and one, a testament to its intricate structure and infinite complexity.

The Cantor set is a powerful example of how a simple process can generate a fractal with remarkable properties. It's a showcase of the power of iteration, the surprising nature of infinity, and the existence of fractional dimensions.

As we continue our journey into the fractal universe, let's remember the lessons of the Cantor set. It shows us that fractals can be found even in the simplest of shapes, that infinity can be infinitely surprising, and that the simple and the complex can coexist in the same figure.

Our next stop will be the world of three-dimensional fractals, where we'll see how fractals can add a new dimension to our understanding of space. But for now let's reflect on the paradoxical world of the Cantor set, its intricate structure, and its mind-bending infinity.

Unfolding Infinity in a Line: The Hilbert Curve

As we continue our exploration of the fractal universe, we encounter yet another fascinating creation: the Hilbert curve. Named after the German mathematician David Hilbert, this fractal curve is a master of deception and a paradox in disguise. It takes us on a mind-bending journey, showing us how a simple line can fill a two-dimensional space.

The Hilbert curve is a space-filling curve, which means that it's a line that can fill a two-dimensional space completely without overlapping or crossing itself. It's a continuous fractal curve that keeps folding onto itself, filling more and more of the space with each iteration until, in the limit, it fills the entire space.

The construction of the Hilbert curve starts with a simple *U* shape. In the next iteration, each straight-line segment of the U-shape is replaced with a scaled-down copy of the original *U* shape. This process continues, with each straight-line segment being replaced with a smaller *U* shape in each subsequent iteration. The result is a complex, infinitely twisting and folding curve that fills a two-dimensional square.

The Hilbert curve is a paradox: it's a one-dimensional line that fills a two-dimensional space. It's a testament to the surprising power of simple iterative processes to generate structures of infinite complexity. It also has practical applications, including in computer science for data storage and retrieval, image compression, and the mapping of multi-dimensional data to one dimension.

The Hilbert curve reminds us that fractals are not just about beautiful patterns and images. They can also be about space, dimension, and the surprising ways in which these concepts can intertwine. It shows us that a simple line can do much more than we might imagine, including filling a two-dimensional space!

As we continue our journey into the fractal universe, let's keep the Hilbert curve in mind. It's a reminder of the power of simple rules, the surprising nature of dimensions, and the infinite complexity that can emerge from iterative processes.

But for now let's pause to appreciate the intricate beauty of the Hilbert curve, its paradoxical nature, and its profound implications for our understanding of space and dimension. In the realm of fractals, the Hilbert curve stands as a testament to the power of simple mathematical rules to generate structures of astounding complexity and beauty.

The Green Delight: Barnsley's Fern

As our journey through the fractal world unfolds, we encounter a fractal that beautifully marries mathematics with nature—the Barnsley fern. Named after the British mathematician Michael Barnsley, this fractal is a stunning example of how simple mathematical rules can generate structures resembling natural forms.

Barnsley's fern is a fractal that uses a method called an iterated function system (IFS) to create a set of points that, when plotted, resemble the delicate, intricate structure of a fern. Barnsley's discovery was revolutionary, because it showed that the complex, organic forms we see in nature could be replicated using simple mathematical processes.

To generate the Barnsley fern, we start with a single point. Then we repeatedly apply one of four transformations to this point, each with a certain probability. These transformations represent the possible ways a new leaf could sprout from an existing one, and the probabilities mirror the likelihood of each sprouting pattern in nature. After thousands or even millions of iterations, the points form the beautiful, leafy shape of a fern.

What's extraordinary about Barnsley's fern is not just its visual similarity to a real fern, but also the simplicity of the mathematical rules that generate it. It's a testament to the hidden mathematical patterns that underlie the natural world and to the power of simple rules to generate structures of remarkable complexity and beauty.

Barnsley's fern also has practical implications. The principles underlying its creation have been used in computer graphics to generate realistic, complex natural forms, from trees and forests to clouds and mountains.

As we continue our journey into the fractal universe, let's remember the lessons of Barnsley's fern. It shows us that the beauty and complexity we see in nature can be understood and replicated through simple mathematical rules. It reminds us that fractals are not just abstract mathematical objects, but also mirrors of the natural world.

But for now let's pause to admire the intricate beauty of Barnsley's fern, its remarkable resemblance to a natural fern, and the profound implications it has for our understanding of the relationship between mathematics and nature. In the world of fractals, Barnsley's fern stands as a testament to the surprising connections between the abstract world of mathematics and the tangible world of nature.

Unleashing the Dragon: The Dragon Curve

As we delve deeper into the fractal world, we discover a fractal that is as captivating as it is mysterious: the dragon curve. This fractal, with its stunning and intricate design, embodies the magic and unpredictability of fractals.

The dragon curve, also known as the Jurassic Park fractal, the Harter-Heighway dragon, or the paper-folding curve, is a fractal that appears like a dragon in flight when plotted. The name "dragon curve" adds an element of fantasy to this mathematical creation, but the formation of this fractal is purely grounded in mathematical rules.

The dragon curve starts with a simple straight line, which is then folded in half over and over again. Imagine you have a long strip of paper. You fold it in half once, then again, and again, each time pressing the crease flat. After you've folded it many times, you unfold it, keeping every crease at a right angle. The resulting shape is the dragon curve.

Each iteration, or folding, creates a more complex pattern, with the curve folding back on itself in an increasingly intricate pattern. The result is a series of zigzagging lines that, when plotted, look like a dragon in flight, hence the name "dragon curve."

The dragon curve is not just a visual spectacle; it's also a mathematical marvel. Despite its apparent complexity, it's generated by a simple process of folding and unfolding, a process that can be described by a simple set of mathematical rules. It's a testament to the power of iterative processes to generate structures of astounding complexity and beauty.

The dragon curve also has practical applications. It has been used in radio antennae design, due to its space-filling properties, in data visualization, and in computer graphics to generate complex shapes and designs.

As we continue our journey into the fractal universe, let's remember the dragon curve. It's a symbol of the power of simple rules and processes, the beauty and complexity that can emerge from them, and the surprising and unexpected forms they can take.

But for now let's pause to admire the intricate beauty of the dragon curve, its remarkable form, and the profound implications it has for our understanding of space, dimension, and complexity. In the world of fractals, the dragon curve stands as a testament to the magic and mystery of mathematical rules and processes, and to the breathtaking structures they can generate.

Voyage into the Third Dimension: The Wonders of 3D Fractals
We've journeyed through the fractal realms of the Sierpiński triangle, Koch snowflake, Mandelbrot set, and Cantor set, exploring the incredible complexity, paradoxes, and beauty that can emerge from simple iterative processes. Now it's time to take a leap into a new dimension as we explore the stunning world of 3D fractals.

Three-dimensional fractals add a whole new level of complexity and beauty to the already astounding world of fractals. They're like the intricate two-dimensional fractals we've seen, but with an extra dimension that adds depth and volume, creating a visual spectacle that is truly mesmerizing.

One of the most famous 3D fractals is the **Menger sponge**. This fractal starts as a simple cube. Like the Sierpiński carpet, it's transformed by repeatedly removing chunks. In the first iteration, the cube is divided into twenty-seven smaller cubes, like a Rubik's cube, and the smaller cubes in the center of each face and the very center of the cube are removed. With each subsequent iteration, the same process is applied to the remaining smaller cubes. The result is an object of infinite detail and complexity, a spongelike figure that is solid yet full of holes.

Another fascinating 3D fractal is the **Julia set in three dimensions**, or **Julia bulb**. Like the 2D Julia set, the 3D Julia bulb is generated by applying a simple rule over and over again. But here the rule involves complex numbers with an added dimension, and the result is a 3D shape of astounding complexity and beauty, filled with swirling, spiraling shapes that resemble bulbs or flowers.

3D fractals challenge and expand our understanding of space, dimension, and complexity. They're a testament to the power of simple rules to generate infinite complexity, even in three dimensions. They show us that fractals are not just flat images or line drawings, but can also be solid, voluminous objects that fill up space in intricate and surprising ways.

As we continue our journey, let's remember the 3D fractals and the new perspective they offer. They remind us that fractals are not just about patterns and repetitions, but also about depth, volume, and the exploration of space.

But for now let's pause and appreciate the intricate beauty of the Menger sponge and the Julia bulb, their complex structures, and their profound implications for our understanding of space and dimension. In the world of fractals, these 3D shapes stand as a testament to the limitless potential of simple mathematical rules to generate breathtaking complexity.

Diving Into Hyperdimensions: Fractals Beyond 3D

As we've journeyed through the fascinating world of fractals, we've encountered shapes that challenge our intuition and push the boundaries of what we understand as dimensions. We've seen how fractals can inhabit the strange territory between our familiar two and three dimensions. But what if I told you that fractals can exist in even higher dimensions, in spaces that we can't directly visualize or experience? Welcome to the world of four-dimensional and higher-dimensional fractals!

First we need some basic ideas to deal with the fourth dimension. Imagine being a cartographer, charting out new territories. With just one dimension, you can map out a straight path—think of a point moving forward or backward along a line. Then you add a second dimension, and you can now navigate a whole plane, going left and right in addition to forward and backward. This is like moving around on a flat map of the world.

Then you discover a third dimension: up and down. Suddenly you're not just navigating a flat map; you're navigating a whole globe. You can move in any direction in three-dimensional space.

Now what if I told you that there was another dimension we could add? Not up, down, left, right, forward, or backward, but something entirely different. That's what we're doing when we start working with **quaternions**.

Quaternions are like an extension of complex numbers. If you remember from your studies, a complex number is a number that has a real part and an imaginary part (like $3 + 4i$). Quaternions take this a step further: they have a real part and three imaginary parts. So a quaternion might look something like this:

$1 + 2i + 3j + 4k$.

"But wait!" you might say. "We know what i is, the imaginary unit. But what are these new things, j and k?"

That's a great question! i, j, and k are the basic units of quaternions. They have a special relationship: each one squared gives -1, and they multiply together in a certain way. For example, if you multiply i and j together, you get k. But unlike regular numbers, the order matters: if you multiply j and i, you don't get k, but $-k$!

To visualize a quaternion, imagine the 3D space you're used to, but with an extra dimension represented by color. So a point in this space isn't just at a certain place left/right, up/down, and forward/backward, but also has a certain color representing the fourth dimension. It's a little abstract, but that's because our brains aren't really wired to think in four dimensions.

In the world of fractals, quaternions help us explore shapes and patterns beyond 3D. They're like our passport to the fourth dimension, letting us discover new mathematical landscapes that we could never reach with real or complex numbers alone.

So as we embark on our journey into higher-dimensional fractals, let's keep these quirky quaternions in mind. They're a bit strange, a bit abstract, but they open a whole new world of mathematical exploration and beauty. After all, who wouldn't want to chart out new territories in the unexplored fourth dimension?

Hyperdimensional fractals are fractals that exist in more than three dimensions. Although we can't visualize these higher-dimensional spaces directly, we can still study and understand them using mathematical tools and representations. And as we might expect from fractals, the results are fascinating.

One example of a higher-dimensional fractal is the **4D Julia set**. We've already encountered the Julia set in two dimensions, but it can also be extended to four dimensions. A 4D Julia set is generated by a simple iterative rule, just like its 2D counterpart. However, instead of iterating a function in the complex plane, we iterate a function in the quaternions, a number system that extends the complex numbers to four dimensions.

Since we can't visualize four dimensions directly, we often visualize 4D Julia sets by taking slices of the 4D space, just like we might take slices of a 3D object to see its internal structure. Each slice is a 3D object, which we can then visualize using standard 3D rendering techniques. The resulting images are incredibly intricate and beautiful, with structures that resemble the 2D Julia set but are even more complex and intricate.

Another example of a higher-dimensional fractal is the **hypersphere**. Just as a 3D sphere is the set of all points in 3D space that are a certain distance from a central point, a hypersphere is the set of all points in a higher-dimensional space that are a certain distance from a central point. While a hypersphere isn't a fractal in the traditional sense, it shares some of the same properties, such as self-similarity. Each slice of a hypersphere is a lower-dimensional sphere, just as each zoomed-in section of a fractal is a scaled-down version of the whole.

These hyper-dimensional fractals and others like them open a whole new world of mathematical exploration. They challenge our intuition, stretch our understanding, and broaden our conception of what a shape can be. They remind us that the world of fractals is not just about beautiful patterns and images, but also about challenging our assumptions, pushing the boundaries of our understanding, and opening new possibilities for exploration and discovery.

So as we continue our journey into the fractal universe, let's keep these higher-dimensional fractals in mind. They're a reminder of the power and the beauty of fractals, the astonishing complexity that can emerge from simple rules, and the endless possibilities for exploration and discovery that they offer. In the realm of fractals, even the sky isn't the limit!

Fractals Around Us: More Common Than You Think!
What if I told you that you've been surrounded by fractals all your life, perhaps without even realizing it? That every time you look out the window, take a walk in the park, or even glance at your own body, you're seeing fractals? It might sound strange, but it's true. Fractals, those beautiful and complex shapes that we've been exploring, are far more common in our everyday world than you might think.

Let's start with something simple: consider a tree. At first glance, it might not seem like anything special. But look a little closer, and you'll start to see a pattern. The trunk splits into large branches, which split into smaller branches, which split into even smaller branches, and so on, down to the smallest twigs. This pattern of branching and rebranching is a classic example of a fractal. It's a pattern that repeats at different scales, creating a complex and beautiful shape out of a simple rule.

But trees are just the beginning. Fractals are everywhere in nature. Think about the veins in a leaf, the pattern of a snowflake, the shape of a mountain range, or the spiral of a seashell. All of these are examples of fractal patterns. They're complex, intricate, and infinitely detailed, but they all come from simple rules and processes.

Even our own bodies are full of fractals. Consider the network of blood vessels in your body. It starts with a few large arteries, which branch into smaller arteries, then into arterioles, and finally into tiny capillaries. This branching network is another fractal, a pattern that repeats at different scales to create a complex and efficient system for distributing blood throughout your body.

Fractals are not just confined to the physical world. They are also prevalent in the realm of sound and music. Have you ever noticed how certain pieces of music seem to have a pattern, a rhythm, that repeats over and over, but each time a little different, a little more complex? That's a fractal pattern. Musicians and composers have been using fractal patterns, perhaps unknowingly, to create intricate and pleasing sounds for centuries.

Even the stock market exhibits fractal behavior! Prices fluctuate in complex patterns that, upon closer inspection, reveal self-similarity over different time scales. A graph of stock prices over a year might look surprisingly similar to a graph of prices over a month or even a week, just at a different scale. Economists and financial analysts use fractal mathematics to model and predict these patterns.

Why are fractals so common? It comes down to efficiency. Fractal patterns allow for maximum complexity and functionality with minimal energy and material use. They provide a way for nature to create intricate structures and patterns out of simple rules and processes.

So the next time you're out for a walk or even just sitting at home, take a moment to look around. See if you can spot the fractals in the world around you. You might be surprised at how common they are and how much beauty and complexity they bring to our everyday lives.

The Unique Properties of Fractals

In our journey thus far, we've seen how fractals manifest themselves in numerous ways both in the natural world and within the realm of mathematics. As we've explored these intricate and mesmerizing forms, it has become apparent that fractals possess some unique properties that set them apart from other geometric shapes and patterns. Let's delve into some of these properties and discover what truly makes a fractal a fractal.

1. Self-similarity: Perhaps the most defining characteristic of fractals is their self-similarity. This means that a fractal pattern, no matter how much you zoom in or out, will always exhibit the same pattern. It's like a never-ending Russian nesting doll. Each part, no matter how minuscule, is a reduced-scale copy of the whole. This property confers on fractals their virtually infinite complexity, as each iteration reveals a new layer of the same pattern.

2. Infinite Detail: Linked closely with self-similarity is the property of infinite detail. Fractals can be zoomed into forever, and with each level of magnification, new details emerge. No matter how closely you look, a fractal will never become a simple straight line or smooth curve. There will always be new twists and turns, bumps and valleys to explore.

3. Noninteger Dimensions: Unlike regular shapes that exist neatly in one, two, or three dimensions, fractals often have noninteger, or fractional dimensions, which is where their name comes from. This strange property arises from their intricate detail and self-similarity. For instance, the Koch snowflake, when examined closely, is more than a one-dimensional line but less than a two-dimensional shape, hence it exists in a fractional or fractal dimension.

4. Simple Rules, Complex Shapes: Despite their complexity, fractals are typically generated by simple rules or equations repeated over and over again. This process, known as iteration, is what generates the intricate and beautiful patterns we associate with fractals. It's a testament to the power of recursion and the surprising complexity that can arise from simple beginnings.

5. Unbounded Perimeter, Bounded Area: Many fractals have the paradoxical property of having an unbounded (infinite) perimeter but a bounded (finite) area. The Koch snowflake is a classic example of this: with each iteration, the perimeter of the snowflake increases, but the area remains finite.

These properties, among others, make fractals a fascinating subject of study. They challenge our traditional understanding of dimensions and spatial organization, and they reveal the inherent beauty and complexity that can arise from simple rules. Fractals are not just mathematical curiosities; they are reminders of the endless wonder and intricacy that can be found in the mathematical universe, and indeed, in the universe at large.

2: WHAT ARE DIMENSIONS AND WHY DO THEY MATTER?

"I can't go back to yesterday because I was a different person then." Presently she began again. "I wonder if I've been changed in the night? Let me think: was I the same when I got up this morning? I almost think I can remember feeling a little different. But if I'm not the same, the next question is 'Who in the world am I?' Ah, that's the great puzzle!"
—Lewis Carroll, *Through the Looking-Glass, and What Alice Found There*

As we embark on the next phase of our fractal journey, it's time to delve into a concept that is integral to understanding these fascinating mathematical structures: dimensions. We're all familiar with dimensions in a general sense. We live in a three-dimensional world, after all. Up and down, left and right, forward and backward; these are the dimensions that govern our everyday existence. But what exactly are dimensions? And why do they matter, especially when we're talking about fractals?

At its most basic, a dimension refers to the minimum number of coordinates needed to specify any point within it. In a one-dimensional space, like a straight line, only one coordinate is needed: the position along the line. In two dimensions, like on a flat surface or plane, two coordinates are required, typically referred to as length and width, or x and y. A three-dimensional space, like the world we inhabit, requires three coordinates, usually expressed as length, width, and height, or x, y, and z.

Dimensions help us understand and navigate the world around us. They give us a framework within which to measure, perceive, and interact with our surroundings. For instance, when you want to meet a friend at a specific location, you provide them with three-dimensional coordinates—a longitude, a latitude, and possibly an altitude or floor level.

However, the world of fractals introduces us to an entirely new perspective on dimensions, one that transcends our everyday, intuitive understanding. Fractals exist in what is often referred to as fractional or noninteger dimensions. This might seem like a contradiction. After all, how can something be half a dimension or 1.6 dimensions? But remember, dimensions in mathematics aren't limited to the tangible, three-dimensional world we're accustomed to.

In the case of fractals, dimensions become a way to quantify the complexity of the object. The more complex and detailed the object, the higher its fractal dimension. A fractal dimension isn't necessarily an integer because it's not describing a volume in a traditional, three-dimensional sense. Instead, it's describing the scaling properties of a fractal—how the detail of the fractal changes with the scale at which it is viewed.

Why does this matter? Because it challenges and expands our understanding of space and dimensions. It opens the door to a world where the lines between dimensions blur, where shapes can exist between our standard dimensions. It's in this in-between world where fractals come to life, and the understanding of dimensions becomes the key to unlocking their secrets.

By understanding dimensions, we are better equipped to explore the infinite complexity of fractals, to appreciate their intricate detail and self-similarity, and to comprehend their seemingly paradoxical existence in fractional dimensions. So hold on tight as we venture deeper into this multidimensional universe of fractals!

Fractals Between Zero and One Dimension

As we journey into the fractional dimensions of fractals, we begin with those that exist between zero and one dimension. The concept of anything existing between a point (zero-dimensional) and a line (one-dimensional) may seem odd at first, but in the world of fractals, it's entirely possible. Let's explore these intriguing mathematical entities, the equations that describe them, and the minds that discovered them.

One of the simplest and most famous examples of a fractal existing between zero and one dimension is the Cantor set, which we introduced in the previous chapter. Named after the German mathematician Georg Cantor, this fractal is constructed by taking a line segment (which is one-dimensional), removing the middle third, then repeating the process with the remaining two segments ad infinitum.

The resulting fractal is a set of points (which are zero-dimensional), but due to the infinite process, it has a dimensionality that is greater than zero but less than one. In fact, the fractal dimension of the Cantor set is approximately 0.63, a strange thought when we consider our conventional understanding of dimensions.

The Cantor set can be described mathematically using recursion, with this equation:

$C\_n = (1/3)C\_(n-1) \cup ((2/3) + (1/3)C\_(n-1))$

where ∪ denotes the union of two sets and n represents the iteration number. The base case is $C\_0 = [0, 1]$, the initial line segment.

Another fascinating fractal in this category is the Koch curve, which we also discussed in the previous chapter. The Koch curve starts as a line segment, but with each iteration, every segment of the line is replaced with a shape that resembles a sideways V. The fractal dimension of the Koch curve is approximately 0.63, the same as the Cantor set, but it looks entirely different, demonstrating the diversity of fractals within the same dimensional space.

The mathematical description of the Koch curve involves a recursive algorithm, and the curve's length increases by a factor of 4/3 with each iteration. The equation for the length after n iterations is

$L\_n = (4/3)^n * L\_0$

where L_0 is the original length.

These discoveries challenged the rigid boundaries of dimensions and opened the door to the concept of fractional dimensions. This was not just a mathematical curiosity, but it had implications for various fields, including physics, computer graphics, and even philosophy, forcing us to rethink our understanding of space, geometry, and dimensions.

The Cantor set and the Koch curve are just two examples of fractals that exist between zero and one dimension. There are many more, each with their own unique properties and mathematical descriptions. As we delve deeper into this fascinating fractal world, we realize that these fractional dimensions are not anomalies but are, in fact, a crucial part of the complex and beautiful mathematical landscape of fractals. And it is in these interstitial spaces, between the whole dimensions we're accustomed to, that some of the most exciting and mind-bending fractal explorations occur. So let's press on as we continue to explore the magical realm of fractals and their dimensions.

From a Point to a Line: Bridging Zero and One Dimension

In our everyday understanding of geometry, the transition from a point (zero dimension) to a line (one dimension) might seem straightforward. However, when we venture into the realm of fractals, this journey becomes an intriguing exploration into the concept of dimensions themselves. So how do we take a point from zero to one dimension in the context of fractals?

Let's start by revisiting our basic understanding of

dimensions. A point, as we know, is zero-dimensional. It has no length, no breadth, and no height. It simply exists, a precise location in space. Now when we extend that point in any one direction, we create a line, which is one-dimensional. The line has length, but no breadth or height. In this sense, we can consider the transition from zero to one dimension as the act of giving length to a point.

In the world of fractals, we can think of this transition as an iterative process. Imagine we start with a point. Now, instead of simply extending this point into a line, we're going to repeat a process that gradually "builds" the point into a fractal line.

Consider the construction of the Cantor set, which we mentioned earlier. We start with a line (one dimension), then remove the middle third of the line, leaving us with two lines, each one-third the length of the original. If we repeat this process with each of the remaining lines, the set of points left behind forms the Cantor set, a fractal with a dimension between zero and one.

Now let's imagine this process in reverse. We could start with a point (zero dimension). In the next step, we create two new points, each a third of the length away from the original point on either side. We now have three points, which we could interpret as a line broken into three segments. If we continue this process, adding two new points around each existing point with each iteration, we would gradually build a fractal line with a dimensionality between zero and one.

This is, of course, a highly simplified and abstracted explanation. The mathematical specifics involve more complex concepts, such as limit sets and recursive algorithms. However, the key takeaway is this: in the world of fractals, dimensions are not rigid and discrete. They can be fractional, and we can transition between them through iterative processes.

This understanding allows us to delve deeper into the complex world of fractals and to explore the beauty and mystery that lies between the dimensions. It's not just about points and lines or surfaces and volumes. It's about a continuum of dimensions, a spectrum of complexity, and a world of infinite detail. So let's continue our journey as we explore more about the magic of dimensions and the world of fractals.

Limit Sets and Recursive Algorithms: Keys to Understanding Fractals

As we journey deeper into the world of fractals, two concepts become increasingly important: limit sets and recursive algorithms. Both are fundamental to understanding the behavior of fractals and the mathematics behind their infinite complexity. In this section we'll explain these concepts in both general and mathematical terms, providing a foundation for anyone wishing to explore the fascinating field of fractal mathematics.

Limit Sets

In general terms, a limit set is a set of points toward which a sequence or a function tends as you repeat the function over and over. In other words, it's the set of "endpoints" of a process when that process is repeated an infinite number of times.

In the world of fractals, limit sets often represent the final form of a fractal when its generating process is repeated indefinitely—for example, the points left in the Cantor set after an infinite number of iterations form the limit set for that process.

Mathematically, limit sets can be defined in terms of sequences or functions. Given a sequence $\{x\_n\}$, the limit set is the set of all points x such that for every positive number ε, there exists a number N (which may depend on ε) such that for all $n > N$, the distance between x_n and x is less than ε. In simple terms, it's the set of all points that the sequence gets arbitrarily close to, and stays close to, as n goes to infinity.

Recursive Algorithms

In general terms, a recursive algorithm is a process that breaks a problem down into smaller, simpler versions of the same problem, then combines the solutions to solve the original problem. It's like a Russian doll, where each step uncovers a smaller doll within, until you reach the smallest doll, which can't be broken down any further.

In the context of fractals, recursive algorithms are used to define and generate the fractal shapes. The algorithm takes a shape (like a line segment in the case of the Koch curve), applies a transformation to that shape (like dividing it into three and replacing the middle segment with two segments to form a V), then repeats that transformation on the new shape.

Mathematically, a recursive algorithm can be defined by a function f that calls itself within its own definition. For example, a simple recursive function in mathematics is the factorial function, denoted by $n!$, which is defined as $n! = n * (n-1)!$, with the base case of $0! = 1$. In the case of fractals, the recursive algorithm often takes the form of a geometric transformation applied to the output of the same transformation.

Understanding these concepts is crucial for anyone interested in studying or working with fractals. They underpin the infinite complexity and self-similarity that define fractals and provide a mathematical framework for generating and analyzing these fascinating shapes. So as we continue our journey into the world of fractals, we'll keep these ideas at the forefront, using them as tools to unlock the secrets of the fractal universe.

Scaling Up: From One Dimension to Two Dimensions

The journey from one dimension to two is a familiar one in our everyday lives. We're used to moving from lines to shapes, from the linear to the planar. But in the world of fractals, this transition becomes a fascinating exploration of dimensionality and infinite complexity. So how do we scale up from one to two dimensions in the context of fractals?

Let's start by revisiting our basic understanding of dimensions. A line, as we know, is one-dimensional. It has length, but no breadth or height. A shape such as a square or a circle, on the other hand, is two-dimensional, with both length and breadth. The transition from one to two dimensions, then, can be seen as the act of giving "breadth" to a "line."

In the fractal realm, this transition can be seen as a process of iterative complexity. We start with a simple, one-dimensional line and then apply a rule or an algorithm that adds a second dimension. This could be as simple as turning a straight line into a zigzag, as we do when creating the Koch curve, or it could involve more complex transformations.

Consider the construction of the Sierpiński triangle, a well-known two-dimensional fractal. We start with a simple equilateral triangle, then remove the middle triangle, leaving us with three smaller triangles. If we repeat this process with each of the remaining triangles, we end up with an intricate fractal shape that fills a two-dimensional plane but still, intriguingly, has a fractional dimension between one and two.

This transition from one to two dimensions isn't just about adding complexity, though. It's also about revealing the hidden depths of the mathematical world. As we move from lines to shapes, from one to two dimensions, we open up new possibilities for exploration and discovery. We see the world in a new light, through the lens of fractal geometry.

So as we continue our journey into the world of fractals, let's embrace this transition from one to two dimensions. Let's celebrate the complexity and the beauty of the shapes we create, and let's marvel at the infinite detail and intricacy that lies between the dimensions—because in the world of fractals, every transition, every transformation, is a step toward a deeper understanding of the universe we live in.

Delving Into the Mathematics of One to Two Dimensions

As we venture from one-dimensional objects to two-dimensional fractals, the mathematics become more involved. In this section, we'll explore the specific mathematical concepts that underlie this transition, focusing on key ideas like self-similarity, recursion, and fractional dimensions.

1. **Self-similarity:** One of the defining characteristics of fractals is their self-similarity. As we scale up from one to two dimensions, we often find that a fractal shape is made up of smaller copies of itself, with each copy scaled down by a constant factor. Mathematically, this can be represented using a similarity transformation or an affine transformation, which preserves the shape's proportions while scaling, rotating, or translating it.

For example, in the Sierpiński triangle, each iteration produces three smaller triangles that are similar to the original triangle, with each smaller triangle having half the side length of its parent triangle. This self-similarity is an essential aspect of the transition from one to two dimensions in the fractal world.

2. **Recursion:** Recursion plays a central role in generating fractals, as we apply a rule or an algorithm repeatedly to create increasingly complex shapes. Mathematically, this can be represented using recursive functions or iterative algorithms that define a sequence of transformations.

In the case of the Koch curve, for instance, we start with a straight-line segment and apply a recursive rule that divides the segment into thirds, replacing the middle third with two segments of the same length and forming an equilateral triangle without the base. By iterating this process, we create a curve that occupies a two-dimensional space, even though it started as a one-dimensional line segment.

3. **Fractional Dimensions:** As we transition from one to two dimensions, we encounter the concept of fractional dimensions. Fractal shapes often have a dimensionality that is not an integer, which is a consequence of their self-similar, recursive nature. This can be quantified using the Hausdorff dimension or the box-counting dimension, both of which provide a measure of the complexity and scaling behavior of a fractal.

For example, the Koch curve has a Hausdorff dimension of log(4)/log(3) ≈ 1.2619, which lies between one and two, indicating that it is more complex than a one-dimensional line but less complex than a two-dimensional shape. This fractional dimensionality is a hallmark of the mathematical richness that emerges as we transition from one to two dimensions in the fractal world.

By delving into the mathematics of the transition from one to two dimensions, we gain a deeper appreciation for the beauty and complexity of fractals. Self-similarity, recursion, and fractional dimensions all play a crucial role in this journey, revealing the intricate patterns and infinite detail that lie at the heart of fractal geometry.

Exploring the Second Dimension: More Than Just a Flat Plane
Let's press on with our interdimensional journey. It's time to explore the second dimension in its full glory, beyond just the confines of flat planes.

The second dimension, at first glance, might seem plain and unadventurous—it's just a flat surface, after all. But when we delve into the world of fractals, we start to see that the second dimension is anything but flat. It's a realm filled with infinite detail, where simple shapes like squares and circles can give birth to a myriad of complex, self-similar forms.

Within the realm of two-dimensional fractals, it's important to distinguish between fractal boundaries and fractal fillings. The boundary of a fractal, such as the outline of the Mandelbrot set or the Koch snowflake, is typically a one-dimensional curve that has a fractional dimension greater than one, reflecting its intricate structure. On the other hand, a fractal filling, like the Sierpiński triangle or the filled Julia set, occupies a two-dimensional area but still has a fractional dimension less than two, due to its self-similar, recursive nature.

As we scale up from one to two dimensions, we see the power of iteration in full display. By repeatedly applying a simple rule or transformation, we can create intricate, two-dimensional fractal shapes that are infinitely complex yet perfectly self-similar. Whether it's the iterative process of adding triangles in the Sierpiński triangle, or the iterative algorithm of the Mandelbrot set that involves complex numbers, iteration is the engine that drives the complexity of two-dimensional fractals.

Finally, our exploration of the second dimension serves as a springboard to higher dimensions. Fractals provide us with a unique tool to visualize and comprehend the complexities of higher-dimensional spaces. For example, the 3D analog of the Sierpiński triangle, known as the Sierpiński tetrahedron or the Sierpiński pyramid, helps us intuitively grasp the concept of a three-dimensional object. Similarly, the study of two-dimensional fractals like the Mandelbrot set and the Julia set paved the way for the discovery of their four-dimensional counterparts in the realm of quaternions.

As we continue our journey from the first dimension to the second, and beyond, we see that the world of fractals is a world of infinite richness and detail. It's a world where simple rules give birth to complex forms, where the flat becomes the intricate, and where the finite unfolds into the infinite. So let's embrace the second dimension in all its fractal beauty, and let's prepare ourselves for the even greater wonders that lie ahead in the higher dimensions.

Ascending from the Plane: The Journey from Two to Three Dimensions

Welcome to the next stage of our dimensional journey, as we ascend from the flat plane of two-dimensional space into the voluminous expanse of three dimensions. This is where fractals start to take on an entirely new level of complexity and fascination.

Exploring the Third Dimension: The step from two to three dimensions can seem like a huge leap, as we move from a flat plane into a voluminous space. Yet the principles that govern the formation of two-dimensional fractals also apply in the third dimension. For instance, the self-similarity and recursive processes that we saw in two-dimensional fractals are equally applicable in the third dimension, leading to the creation of stunningly intricate and complex three-dimensional fractal shapes.

The Magic of 3D Fractals: One of the most remarkable aspects of three-dimensional fractals is their ability to capture our imagination with their complexity and beauty. Shapes like the Mandelbulb, the Menger sponge, and the Sierpiński tetrahedron are mesmerizing to behold, exhibiting an almost otherworldly quality. These shapes fill space in unusual ways, often creating intricate networks of holes and tunnels that reveal more and more detail as we zoom in closer.

The Mathematical Machinery: To understand how we create these three-dimensional fractals, we'll need to revisit some of the mathematical concepts we've explored so far and introduce a few new ones. Techniques such as iteration and recursion are still at the core of generating three-dimensional fractals, but we'll also need to delve into the world of complex numbers and quaternions to fully capture the magic of the third dimension.

The Gateway to Higher Dimensions: Once we've grasped the principles of three-dimensional fractals, we'll be poised to venture even further, into the realm of four-dimensional fractals and beyond. Just as the second dimension served as a springboard to the third, so too does the third dimension prepare us for the intricacies of higher-dimensional spaces.

As we make this ascent from the plane to the third dimension, we'll encounter a world of breathtaking complexity and beauty. It's a world where familiar shapes are transformed into objects of unfathomable intricacy, where space is filled in ways that defy our everyday intuitions, and where the simple rules of fractal geometry generate forms of stunning complexity. So let's buckle up for this next leg of our interdimensional journey, as we explore the fascinating realm of three-dimensional fractals.

Techniques for Creating 3D Fractals: Iteration and Recursion
As we venture into the intriguing world of 3D fractals, it's essential to understand the core techniques that facilitate their creation. In this part, we'll delve into the fundamental roles of iteration and recursion in generating these mesmerizing shapes.

Iteration, as we've already discovered, involves the repeated application of a rule or a function. In the context of 3D fractals, it's this repetitive process that gives rise to the bewildering complexity and infinite detail that these shapes exhibit. Whether we're generating a Menger sponge or a Mandelbulb, the principle remains the same: start with a simple shape, apply a rule repeatedly, and watch as complexity emerges from simplicity.

Closely related to iteration is the concept of recursion. A recursive process is one that feeds into itself, using the output from one stage as the input for the next. When applied to fractal generation, recursion leads to the property of self-similarity, where the whole has the same shape as one or more of the parts. In 3D fractals, this self-similarity manifests in multiple dimensions, resulting in shapes that exhibit the same patterns, whether you're viewing them at a distance or up close.

A perfect example of iteration and recursion at work in 3D fractals is the Menger sponge. The creation of a Menger sponge starts with a cube. The cube is divided into twenty-seven smaller cubes (like a Rubik's cube), and the smaller cube at the center of each face, along with the smaller cube at the very center of the larger cube, are removed. This process is then recursively applied to each of the remaining smaller cubes, over and over again. The result, after an infinite number of iterations, is the Menger sponge, a 3D fractal with an infinite surface area and zero volume!

The magic of 3D fractals lies in the interplay between iteration and recursion. The iterative process adds detail and complexity at each step, while the recursive process ensures that the same patterns repeat at different scales, giving the fractal its characteristic self-similar structure.

As we explore further into the world of 3D fractals, we'll see that these principles of iteration and recursion, so simple at their core, are the driving forces behind some of the most stunning and complex shapes known to mathematics. They show us that sometimes, complexity isn't a result of complicated rules, but a product of simple rules applied repeatedly over time.

Learning the Techniques: A Hands-On Approach to Fractals
As we explore the world of fractals, it's not enough to just learn about them theoretically. To truly appreciate their beauty and complexity, we need to get our hands dirty and dive into the mathematics that generate them. This hands-on approach not only deepens our understanding, but also provides an exciting opportunity to create and explore our own fractal landscapes.

Consider the fundamental technique of iteration, the act of repeatedly applying a rule or function. As we've seen, iteration is central to the formation of fractals. But how does it work in practice? Imagine we're creating the Sierpiński triangle. We start with a simple equilateral triangle. We then apply a rule: divide the triangle into four smaller equilateral triangles by connecting the midpoints of each side and remove the central triangle. This rule is applied repeatedly, each time to the smaller triangles that remain. As we iterate, the pattern becomes more complex, but it always retains its self-similarity, the defining characteristic of fractals.

Now let's turn our attention to recursion. Remember, a recursive process is one that feeds into itself, using the output from one stage as the input for the next. This is beautifully illustrated by the creation of the Koch snowflake.

To create a Koch snowflake, we start with an equilateral triangle, which we'll call the initiator. We then apply a rule to each line segment, or generator. The rule is to divide the generator into three equal parts and replace the middle part with two line segments of the same length, forming an equilateral triangle pointing outward. The process is repeated recursively on each new line segment that is created.

What's fascinating about the Koch snowflake is that as the process continues, the perimeter of the snowflake becomes infinitely long, yet it always encloses a finite area!

Experimenting with Fractals

Now that we have a grasp of the iterative and recursive processes that form fractals, it's time to experiment. With today's technology, we have a range of computer software available that allows us to create our own fractal shapes. Programs like Fractal Explorer, Ultra Fractal, and Mandelbrot Viewer are user-friendly tools that allow us to input simple rules and observe the complex output. By experimenting with different rules and observing the results, we can gain a deeper appreciation for the profound beauty and complexity that fractals offer.

As we continue our journey into the world of fractals, remember that these shapes are more than just mathematical curiosities. They are a testament to the power of simple rules, iteratively and recursively applied, to generate infinite complexity. And perhaps they offer a new way to perceive and understand the intricate patterns that underlie our universe.

Having Fun with Ultra Fractal: Your Personal Fractal Playground

Let's dive into the practical side of fractals and engage in a thrilling, hands-on adventure with Ultra Fractal, a powerful and user-friendly software that lets you create your own mesmerizing fractal art.

Setting Up Your Adventure: First things first; you need to download Ultra Fractal. It's available for both Windows and Mac OS. Once it's downloaded and installed, you'll see a simple interface with a main window displaying a default fractal image and several smaller windows with different options and parameters.

The main window is where the magic happens. It displays

the fractal generated based on the parameters you set in the other windows. You can zoom in and out by simply clicking and dragging your mouse. You'll quickly notice the infinite complexity and self-similarity of fractals. No matter how much you zoom in, you'll always find new and intricate details to explore!

The smaller windows contain a treasure trove of options and settings that allow you to manipulate the fractal image. You can choose different fractal types from the Formula tab, adjust color gradients, layer multiple fractals together, and much more. It may feel overwhelming at first, but don't worry. The beauty of Ultra Fractal is that it's as simple or as complex as you want it to be. Start with simple changes and gradually experiment with more complex settings as you gain confidence.

Once you're familiar with the software, it's time to unleash your creativity! Try out different formulas, tweak the parameters, layer fractals, adjust the colors, and watch as your fractal morphs into a unique work of art. Remember, there are no rules or restrictions in this playground. The only limit is your imagination!

Ultra Fractal makes it easy to save your fractal creations and share them with the world. You can save your fractals as images in various formats, or even as animations if you're using the Extended Edition of the software. Exploring fractals with Ultra Fractal is a fun and exciting way to interact with these fascinating mathematical structures. It's a journey of discovery that never ends, as each new parameter and formula reveals a whole new fractal landscape to explore. So let your curiosity guide you, and enjoy the ride!

Exploring the Mandelbrot Set with Mandelbrot Viewer: Your Guide to the Infinite Universe

The Mandelbrot set, with its infinite complexity and mesmerizing beauty, is one of the most famous fractals. Let's take a personal tour with the help of Mandelbrot Viewer, a user-friendly software that allows you to explore this fascinating fractal universe.

1. **Launching the Mission:** First off, download and install Mandelbrot Viewer from their official website. Once it's launched, you'll be greeted with a vivid display of the Mandelbrot set.

2. **Steering the Spaceship:** Navigating Mandelbrot Viewer is straightforward. The main interface displays the Mandelbrot set, and you can zoom in and out using your mouse wheel or the zoom buttons. Clicking and dragging allows you to move around the set. You'll quickly notice the self-similarity property of fractals—no matter how much you zoom in, the Mandelbrot set continues to reveal smaller copies of itself with slight variations.

3. **Adjusting the Lens:** The software also allows you to change the color scheme, providing various ways to visualize the Mandelbrot set. The Color menu gives you options to invert colors, choose from preset color themes, or even create your own themes. Play around with these settings to see the Mandelbrot set in a different light!

4. **Going Deeper:** One of the unique features of Mandelbrot Viewer is its Julia Mode. By clicking on a point in the Mandelbrot set, you can generate the corresponding Julia set, which is another type of fractal that's intimately related to the Mandelbrot set. This feature provides a deeper understanding of the relationship between these two iconic fractals.

5. **Recording Your Journey:** You can save any interesting views you find during your exploration. Simply go to the File menu and choose Save Image. You can choose different formats for your image and even decide the resolution. This feature allows you to share your discoveries with others or just keep a personal record of your journey into the infinite world of the Mandelbrot set.

Exploring the Mandelbrot set with Mandelbrot Viewer is like embarking on a never-ending journey into a world of infinite complexity and beauty. So buckle up and enjoy the ride!

Unleashing Your Creativity with Fractal Lab: A Beginner's Guide
Fractal Lab is a fantastic tool that blends art and science in an interactive 3D fractal creator. It allows users to delve into the depths of the fractal universe and construct their own fractal landscapes. Here's a beginner's guide to help you get started on your Fractal Lab adventure.

1. **Launching the Lab:** Head over to the Fractal Lab website and click the Launch Fractal Lab button. The application runs in your Web browser, so there's no need to download or install anything. You'll be presented with a default 3D fractal, which you can start exploring right away.

2. **Manipulating the Masterpiece:** The main window in Fractal Lab displays your current fractal. You can zoom in and out, rotate, and pan the fractal using your mouse or touchpad. Experiment with these controls and watch as your fractal reveals its intricate details from every angle.

3. **Tweaking the Tool**s: On the right side of the screen, you'll find a panel with a variety of parameters you can adjust to change the shape, color, lighting, and texture of your fractal. Each parameter slider affects a different aspect of the fractal, and even slight adjustments can produce dramatic changes. Don't be afraid to experiment and see what happens!

4. **Saving and Sharing:** Fractal Lab provides options to save your creations and even share them with others. You can save your fractal parameters for future use, export your fractal as an image, or create a URL that links directly to your current fractal view.

5. **Diving Deepe**r: Once you've got the hang of the basics, there's plenty more to explore. Fractal Lab supports a wide range of fractal types, including Mandelbrot, Julia, and many more. The Fractal tab in the parameters panel lets you choose different fractal formulas and adjust their parameters.

Using Fractal Lab is a unique and rewarding experience that combines creativity, exploration, and learning. It's a tool that allows you to not just observe fractals, but interact with them, manipulate them, and create your own unique fractal universes. So dive in and let your creativity run wild!

From 3D to 4D: Fractals in Higher Dimensions
The last station on our dimensional journey is perhaps the most exciting and mysterious one. It's a leap from the comfortable realm of 3D that we all know and love into the uncharted territory of the fourth dimension. Like a door opening to an entirely new world, the fourth dimension unfolds a world of fractals that is even more complex and beautiful than the one we've already explored.

The very mention of the fourth dimension may seem like an abstract, arcane concept, and you might be wondering: what does it even mean? In our everyday life, we navigate through three dimensions: we move left or right, forward or backward, up or down. But mathematically, we can venture beyond this three-dimensional framework. A fourth dimension can be conceived as an additional direction, perpendicular to the three we're familiar with.

Though we can't physically perceive the fourth dimension, we can attempt to understand it through analogies. Imagine a two-dimensional creature living on a flat surface, unaware of the concept of up or down. How could we explain a cube, a three-dimensional object, to this creature? By slicing the cube into two-dimensional cross-sections, the creature would start to grasp the essence of the cube. Similarly, we can "slice" 4D shapes into 3D sections we are able to visualize.

Now, let's venture into the mesmerizing world of 4D fractals. Just like our journey from 2D to 3D fractals, we find that familiar fractals have their 4D counterparts. One notable example is the Mandelbulb, the four-dimensional brother of the Mandelbrot set. It retains the self-similarity and infinite complexity that characterizes the Mandelbrot set, but the additional dimension introduces a whole new level of intricacy.

Exploring 4D fractals may feel like stepping into a fantastical world where familiar rules are bent and new laws of nature take hold. These fractals are not just mathematical curiosities. They represent the capacity of the human mind to venture beyond the boundaries of our physical world, to imagine and explore spaces that we cannot see or touch but can nonetheless comprehend.

So let's keep our minds open as we continue to journey deeper into the wonderful world of fractals. Remember, in this realm, dimensions are not barriers, but invitations to explore further, to understand more, and to marvel at the beauty of mathematical structures that are as complex and fascinating as the universe we inhabit.

From 3D to 4D: Diving Deeper

Having understood the concept of the fourth dimension and its significance, let's now explore some specific 4D fractals, starting with the 4D analogue of the Mandelbrot set, called the Mandelbulb.

The Mandelbulb is generated by a simple, iterative formula, just like the Mandelbrot set. However, instead of squaring the complex number at each step, as we do for the Mandelbrot set, we raise it to the power of eight. This introduces an additional degree of complexity that manifests as a fourth dimension.

To understand this, let's imagine we're working with a number of the form $x + yi + zj + wk$, where i, j, and k are the units of the imaginary j, and k dimensions, and x, y, z, and w are real numbers. Squaring this number would result in a four-dimensional result (since $i^2 = j^2 = k^2 = ijk = -1$), which we can't easily visualize. Instead, we can consider a series of three-dimensional "slices" of the resulting four-dimensional shape, much like a doctor viewing a series of 2D X-ray images to understand the 3D structure of the human body.

Similarly, we could explore the Julia set in four dimensions. A 4D Julia set is also generated by iterating a formula, but with a slight twist. The formula is a function of the coordinates of each point in four-dimensional space, and these coordinates are fed back into the function at each iteration. This creates a fractal that is not self-similar in the usual sense but exhibits a kind of four-dimensional symmetry that is equally mesmerizing.

These are just a few examples of the fascinating world of 4D fractals. By understanding the principles behind them, you are well equipped to delve into more complex fractals and even create your own. Remember, the beauty of fractals lies not just in their visual appeal, but also in the intriguing mathematical principles they embody. Happy exploring!

Visualizing and Creating 4D Fractals

To truly grasp the concept of four-dimensional fractals, one must engage in a bit of creative, out-of-the-box thinking. This involves not only understanding the math behind these objects, but also finding innovative ways to represent them visually.

One popular method for visualizing 4D fractals is through "dimensional cross-sections." Imagine cutting a three-dimensional object with a two-dimensional plane: you'd end up with a two-dimensional cross-section of that object. Similarly, if we were to slice a four-dimensional object with a three-dimensional plane, we'd end up with a three-dimensional cross-section of that 4D object. This is the concept behind creating 3D renditions of 4D fractals, such as the Mandelbulb.

However, a more immersive method involves animation. By changing the position of the 3D cutting plane over time, we can create the illusion of moving through the fourth dimension. This approach has been used to create breathtaking animations of four-dimensional fractals, which provide a more dynamic and intuitive understanding of these complex shapes.

As for creating your own 4D fractals, there are several tools available. For instance, a software called Fragmentarium is a fantastic platform for experimenting with fractals in higher dimensions. It's an open-source, cross-platform IDE for exploring pixel-based graphics on the GPU. It is a simple way to create complex images, including 4D fractals.

Another powerful tool is Mandelbulber, which supports a variety of formulas for generating 3D fractals, including the Mandelbulb and Julia set. Mandelbulber also allows for the exploration of fractals in higher dimensions. By adjusting parameters, rotating, and zooming, you can explore these fractals and create stunning images and animations.

These are just a couple of examples of the tools available for creating and visualizing 4D fractals. With a bit of practice and experimentation, you can start creating your own beautiful, complex fractals and see firsthand the intricate patterns and shapes that emerge from the simple rules of iteration and recursion.

Remember, the key to understanding and appreciating fractals lies in exploration and curiosity. Don't be afraid to experiment, make mistakes, and ask questions. The world of fractals is vast and beautiful, and there's always something new to discover.

Using Mandelbulber to Create 3D and 4D Fractals

Mandelbulber is an exciting tool that lets you delve into the world of fractals, creating stunning images in three and four dimensions. Here's a general guide to help you get started. Remember, the beauty of Mandelbulber lies in its ability to let you experiment and play, so don't be afraid to tweak settings and see what happens!

Getting Started: First, download and install Mandelbulber from the official website. Once you open the software, you'll be greeted with a 3D view of a fractal and a complex-looking interface full of sliders, buttons, and menus. Don't be overwhelmed; we'll break it down.

Understanding the Interface: The interface is divided into several panels. The most important ones are the Parameters panel, where you can adjust the properties of the fractal, and the Render panel, where you can control the quality of the final image.

Creating a 3D Fractal

- **Selecting a Fractal Type:** In the Fractal tab of the Parameters panel, you can select the type of fractal you want to generate. Let's start with the 3D Mandelbulb.
- **Adjusting the Fractal Parameters:** Depending on the fractal type, there will be different parameters you can adjust to change the shape of the fractal. For the 3D Mandelbulb, you can adjust the Power slider to change the complexity of the fractal.
- **Navigating the 3D View:** Use the mouse to rotate the view and the WASD keys to move around. You can also use the scroll wheel to zoom in and out.
- **Rendering the Fractal:** Once you're happy with the view, go to the Render panel and click the Render button. Depending on your settings, this may take a few seconds to a few minutes.

Creating a 4D Fractal: Creating a 4D fractal involves a few extra steps:

- **Selecting a 4D Fractal Type:** In the Fractal tab, select a fractal type that supports 4D, such as the Quaternion fractal.
- **Adjusting the 4D Parameters:** In addition to the regular fractal parameters, there will be additional parameters for controlling the fourth dimension. These can be found in the 4D tab of the Parameters panel. For example, the W slider adjusts the position of the 3D cutting plane in the fourth dimension.
- **Animating the 4D Fractal:** To create the illusion of moving through the fourth dimension, you can animate the W parameter. To do this, go to the Animation panel and set the Start and End values for the W parameter. Then, click the Render Animation button. This will create a series of images that, when played back, create the illusion of motion through the fourth dimension.

Remember, Mandelbulber is a complex tool with many features to explore, so don't be afraid to experiment and see what you can create!

Playing with 4D Toys to Understand and Create 4D Fractals

4D Toys is a fascinating tool that not only introduces the concept of a fourth spatial dimension but also lets you play with it in an immersive environment. This interactive "toy box" filled with four-dimensional shapes aims to provide an intuitive understanding of this often-confusing dimension.

While 4D Toys does not specifically create fractals, it does offer a unique playground for experimenting with and understanding the fourth dimension, which could, in turn, aid in grasping the concept of four-dimensional fractals.

Getting Started: Download 4D Toys from the official website or the app store if you're using a tablet. Once you open the program, you'll find a play area with several objects and a toy box filled with more shapes.

Understanding the Interface: The interface is simple and intuitive. The play area is where you interact with objects, and the toy box, in the corner, contains different 4D objects to choose from. There's also a slider that lets you adjust your viewpoint in the fourth dimension.

Interacting with 4D Objects
- **Selecting Objects:** Click or tap on an object in the toy box to select it. You can then click or tap anywhere in the play area to place the object.
- **Manipulating Objects:** Use your mouse or finger to move objects around. You can also rotate objects in 4D by right-clicking and dragging (or using a two-finger drag on a touch screen).
- **Adjusting the Viewpoint:** Use the slider to change your viewpoint in the fourth dimension. This can be a bit disorienting at first, but with practice, you'll start to get an intuition for how objects behave in 4D.

Understanding 4D Through Play

As you play with the objects, you'll notice some strange phenomena. For example, objects can pass through each other without colliding, and objects can turn inside out without breaking apart. These behaviors are natural in four dimensions, even though they seem impossible in our three-dimensional world.

While you can't create 4D fractals directly in 4D Toys, playing with the program and observing how objects behave in four dimensions can help you better understand the concept of four-dimensional fractals. By visualizing how shapes can exist and interact in this higher dimension, you'll be better equipped to grasp the idea of fractals that extend into the fourth dimension.

Remember, 4D Toys is designed for exploration and play, so don't be afraid to experiment and see what you can discover!

The Mathematical Side of Fractal Dimensions

Understanding the dimension of a fractal requires a shift in perspective from our traditional understanding of dimensions. In the classical sense, a line is one-dimensional, a square is two-dimensional, and a cube is three-dimensional. However, fractals complicate this notion because they exhibit complexity and detail at every scale, causing them to occupy a space between these standard dimensions. This property is encapsulated in the concept of the fractal dimension.

The fractal dimension, also known as the Hausdorff dimension, isn't an integer. It's a number that often falls between the standard dimensions, giving us insight into the fractal's intricate structure. To calculate this, we employ a specific formula that measures how the detail of a fractal changes with the scale at which it is observed.

To understand this, let's use the Sierpiński triangle as an example. It's a fractal that starts with an equilateral triangle. With each iteration, we remove the middle triangle of the remaining larger triangles, which results in creating smaller triangles ad infinitum.

Let's define a ratio (r) as the side length of a small triangle to the side length of the large triangle. In the Sierpiński triangle, r is 1/2. Let's also define a number (N) as how many smaller copies fit into the original shape. In this case, N is 3. The formula for the fractal dimension (D) is given by the following:

$D = -\log(N) / \log(r)$.

If we plug in the values for the Sierpiński triangle, we get this:

$D = -\log(3) / \log(1/2) \approx 1.585$.

This tells us that the Sierpiński triangle exists somewhere between one and two dimensions, which matches our intuitive understanding of this fractal. It's more than a line (1D) but less than a plane (2D). By calculating the fractal dimension, we can quantify a fractal's complexity and self-similarity. This method works with other fractals as well, although the calculations can become more complicated depending on the nature of the fractal.

Understanding the mathematical basis of fractal dimensions allows us to appreciate the complexity and beauty of fractals even more. It also opens the door to further exploration of these fascinating shapes, their properties, and their wide-ranging applications.

Calculating Fractal Dimensions: More Examples

Now that we've covered the basic idea of calculating a fractal's dimension, let's examine a couple more examples to solidify our understanding.

Koch Snowflake: The Koch snowflake is a fractal that starts with an equilateral triangle. With each iteration, we add an equilateral triangle one-third the size to each side, creating a starlike shape. This process continues infinitely, adding more detail to the snowflake with each iteration.

The ratio (r) in this case is 1/3, as each new triangle added is a third the size of the section it's added to. The number of self-similar copies (N) is 4, as each line segment is divided into four parts.

Let's plug these into our formula:
$D = -\log(N)/\log(r)$,
$D = -\log(4)/\log(1/3) \approx 1.261$.

So the Koch snowflake has a dimension slightly above one, meaning it's more complex than a line but less than a plane.

Cantor Set: The Cantor set is a fractal that starts with a line segment. Each step involves removing the middle third of each segment, which results in two smaller segments. This process is repeated indefinitely with each of the remaining segments.

In the Cantor set, the ratio (r) is 1/3, because we're cutting each segment into thirds, and the number of self-similar copies (N) is 2, because we end up with two segments after each cut.

Plugging these into the formula gives us this:
$D = -\log(N)/\log(r)$,
$D = -\log(2)/\log(1/3) \approx 0.631$.

This indicates that the Cantor set is a fractal that exists in a dimensional space between a point (0D) and a line (1D).

As you can see, the fractal dimension gives us a precise way to talk about the complexity and detail of fractals in a way that the standard concept of dimension can't capture. While the math can seem daunting at first, with practice, it becomes a powerful tool for exploring the intricate and beautiful world of fractals.

Calculating Dimensions of Higher-Dimensional Fractals

When dealing with higher-dimensional fractals, the basic principle remains the same. The dimension is calculated based on the ratio of sizes and the number of self-similar copies. However, the process can be more complex due to the increased dimensionality. Let's delve into a few examples.

Menger Sponge: The Menger sponge is a three-dimensional fractal that starts with a cube. In the first iteration, we divide each face of the cube into nine squares (like a Rubik's cube), then remove the central square from each face and the cube at the center, leaving twenty smaller cubes.

In this case, the ratio (r) is 1/3, as each new cube added is a third the size of the cube it's added to. The number of self-similar copies (N) is 20. Plugging these into our formula gives us this:

$D = -\log(N) / \log(r)$,

$D = -\log(20) / \log(1/3) \approx 2.727$.

So despite the Menger sponge being built from 3D cubes, its fractal dimension is less than three.

Sierpiński Tetrahedron (3D Pyramid): The Sierpiński tetrahedron, or tetrix, starts with a solid tetrahedron. For each iteration, we shrink the original tetrahedron to half its edge length and create four self-similar copies, which are placed at the corners of the original tetrahedron.

Here, the ratio (r) is 1/2 (each new tetrahedron is half the size of the previous one), and the number of self-similar copies (N) is 4.

We apply the formula:

$D = -\log(N) / \log(r)$,

$D = -\log(4) / \log(1/2) \approx 1.585$.

This result indicates that the Sierpiński tetrahedron, despite being formed from 3D tetrahedrons, has a fractal dimension between one and two.

With these examples, we can see how the concept of fractal dimension allows us to understand the complexity of higher-dimensional fractals in a quantitative way. It's a remarkable tool that bridges the gap between intuition and mathematical formalism, helping us to better understand the intricate nature of fractals.

Soaring Into Even Higher Dimensions

The concept of fractal dimensions extends beyond the third dimension. While it becomes increasingly challenging to visualize fractals in higher dimensions, the mathematical principles remain the same. Let's explore a few examples.

4D Sierpiński Tetrahedron: Let's consider a four-dimensional analogue of the Sierpiński tetrahedron, which we'll call a

Sierpiński simplex. In the first iteration, we start with a four-dimensional hyperpyramid (also known as a four-simplex), which is a 4D shape bounded by five tetrahedra. We then remove the central pyramid and the four pyramids that connect the central pyramid to the vertices, leaving five smaller hyperpyramids.

In this case, the ratio (r) is 1/2 (each new hyperpyramid is half the size of the previous one), and the number of self-similar copies (N) is 5. Using our formula, we get this:

$D = -\log(N) / \log(r)$,
$D = -\log(5) / \log(1/2) \approx 1.161$.

Even though the Sierpiński simplex is formed from 4D hyperpyramids, its fractal dimension is a little over one.

Higher-Dimensional Menger Sponge: For the higher-dimensional Menger sponge, we start with a hypercube in n-dimensions. In the first iteration, we divide each face into (n-1)-dimensional cubes and remove the central cube from each face, along with the central hypercube. The process is similar to that of the 3D Menger sponge but extended into higher dimensions.

The ratio (r) is 1/3, as each new cube is a third the size of the cube it's added to. The number of self-similar copies (N) is 3^{n-1}.

For a 4D Menger sponge, N would be $3^{(4-1)} = 27$. The fractal dimension (D) would be this:

$D = -\log(N) / \log(r)$
$D = -\log(27) / \log(1/3) \approx 3.631$

***

These examples illustrate that the concept of fractal dimension continues to hold value in higher dimensions. Despite the difficulty in visualizing these shapes, the mathematical principles offer a bridge between our three-dimensional intuition and the reality of higher-dimensional space. In the realm of fractals, dimensions become a fluid concept, paving the way for a deeper understanding of the universe's complex patterns.

Exploring Dimensions: Your Fractal Quest
Below are several problems that will help you explore the world of fractals and their dimensions using the programs Mandelbulber and Ultra Fractal.

- **Problem 1: Exploring Fractals between Zero and One Dimension**

  Using Ultra Fractal, create a fractal line that exists between zero and one dimension. Start with a line and use the program to divide it into segments, creating a fractal. Document the steps you took and explain how the division process contributed to creating a fractal of less than one dimension.

- **Problem 2: From 1D to 2D with Sierpiński's Triangle**

  Use Ultra Fractal to generate a Sierpiński triangle. Start by creating an equilateral triangle and iteratively remove smaller triangles. Write down the steps you took to create the fractal and explain how each iteration brought you closer to a two-dimensional object from a one-dimensional line.

- **Problem 3: Journey from 2D to 3D with the Menger Sponge**

  With Mandelbulber, create a Menger sponge. Begin with a cube and iteratively remove smaller cubes. Document your steps and describe how the transition from two dimensions to three dimensions occurred.

- **Problem 4: Visualizing the Fourth Dimension with a Hypercube**

  Using Mandelbulber, generate a four-dimensional hypercube (also known as a tesseract). As this is a challenging task, start by understanding how a cube can be generated from squares in 3D space. Then, try to apply this knowledge to generate a 4D cube from cubes. Record your steps and describe how you visualized the transition from 3D to 4D.

- **Problem 5: Soaring Into Even Higher Dimensions**

  Finally, create a fractal that exists in more than four dimensions. This will be a theoretical exercise, as visualizing such a fractal would be nearly impossible. However, you can use the mathematical principles discussed in this chapter to describe the fractal's properties and how it would be generated.

Remember, fractals stretch the boundaries of our understanding of dimensions, and working with them is more about the journey than the destination. Don't worry if you can't visualize these higher-dimensional fractals; the goal is to immerse yourself in the process and deepen your understanding of the infinite complexity of fractals.

Here are the explanations to the problems we just visited:

- **Problem 1: Creating a Koch Curve**

A Koch curve is a fractal that begins as an equilateral triangle. Each straight line in the shape is divided into three sections of equal length. The middle section is then replaced with two lines of the same length, forming an equilateral bump. This process is repeated indefinitely for each line segment in the shape, leading to the fractal known as the Koch curve.

For each iteration, the length of the curve increases by a factor of 4/3. Therefore, as the number of iterations approaches infinity, the length of the curve tends to infinity as well. Despite this, the area enclosed by the Koch curve remains finite. This property is a hallmark of fractals.

- **Problem 2: Creating a Sierpiński Triangle**

A Sierpiński triangle starts as a single equilateral triangle. Then the triangle is divided into four smaller equilateral triangles, and the middle one is removed. This process is repeated for each of the remaining smaller triangles, and so on, indefinitely.

With each iteration, the area of the Sierpiński triangle decreases, while the perimeter increases. As the number of iterations approaches infinity, the area tends to zero, but the perimeter tends to infinity. This is another classic characteristic of fractals.

- **Problem 3: Creating a Menger Sponge**

A Menger sponge starts as a cube. The cube is divided into twenty-seven smaller cubes (like a Rubik's cube), and the central cube in each face, as well as the cube in the center, are removed.

This process is repeated for each of the remaining smaller cubes, indefinitely.

The Menger sponge is a three-dimensional fractal. As the number of iterations approaches infinity, the volume of the Menger sponge tends to zero, while its surface area tends to infinity.

- **Problem 4: Creating a 4D Fractal**

Fractals in 4D are more complex to visualize and understand. Essentially, these shapes extend the principles of fractals into the fourth dimension, often conceptualized as time or an additional spatial dimension.

A 4D fractal could start as a 4D shape (like a tesseract, the 4D analog of a cube), and apply a similar process of division and removal as in the previous examples, but in four dimensions.

As with the previous examples, as the number of iterations approaches infinity, the 4D volume (hypervolume) tends to zero, while the 4D surface area (three-dimensional boundary) tends to infinity.

These are generalized answers, and the specifics could vary depending on the exact problems and the fractals involved.

- **Problem 5: Creating a Fractal in Higher Dimensions**

Once you've grasped the concept of fractals in lower dimensions and the fourth dimension, the generalization to higher dimensions follows the same principles.

Let's consider a five-dimensional fractal. As a base object, we could take a penteract, which is the 5D analogue of a cube. Similar to the Menger sponge, divide the penteract into 5^5 smaller penteracts and remove the central ones along each 4D face as well as the one in the center.

This process is repeated for each of the remaining smaller penteracts, indefinitely. As with the previous examples, as the number of iterations approaches infinity, the 5D volume (5D hypervolume) of this fractal tends to zero, while the 5D surface area (four-dimensional boundary) tends to infinity.

This is, of course, very abstract and nearly impossible to visualize. However, the mathematical principles remain consistent, regardless of the number of dimensions involved. This is part of the beauty and complexity of fractals—their self-similar nature allows them to be described and understood even in dimensions we can't physically observe or intuitively comprehend.

How to Use Ultra Fractal to Find the Answers

Problem 1: Ultra Fractal

To create a Koch curve in Ultra Fractal:
1. Download and open Ultra Fractal.
2. Go to File and then New to create a new fractal.
3. Choose Standard.ufm from the Fractal Mode drop-down menu, and Koch from the Fractal Type drop-down menu.
4. Click on the Layers tab, and then click on Outside.
5. Set the transfer function to None.
6. Click on Inside and set the transfer function to Solid.
7. Adjust the color and depth as desired.

Problem 2: Ultra Fractal

To create a Sierpiński triangle in Ultra Fractal:
1. Follow the same steps as above to create a new fractal.
2. This time, select Sierpiński from the Fractal Type drop-down menu.
3. Adjust the color and depth as desired.

Problem 3: Mandelbulber

To create a Menger sponge in Mandelbulber:
1. Download and open Mandelbulber.
2. Go to File and then New to create a new fractal.
3. Choose Menger3 from the Fractal Formula drop-down menu.
4. Adjust the parameters as desired to explore the three-dimensional nature of the Menger sponge.
5. Use the navigation controls to move around your 3D fractal.

Problem 4: Mandelbulber

To create a 4D fractal in Mandelbulber:
1. Follow the same steps as above to create a new fractal.
2. Choose Hypercomplex from the Fractal Formula drop-down menu. This formula introduces a 4D component to the fractal.
3. Adjust the parameters as desired to explore the 4D nature of the fractal.

***

Remember that exploring higher dimensions visually can be challenging, but these tools can provide a sense of the complexity and beauty of these mathematical objects. Experiment with different parameters and settings to create unique fractals and deepen your understanding of these concepts.

3: GRASPING INFINITY: THE PARADOX OF FRACTALS

That which the school of Zeno has striven to say, that "everything is one," Heraclitus seems to me to have said more clearly than anyone else. And he is better able than they to persuade the multitude; for the many believe that opposites are opposed to each other and do not understand that they are in harmony with each other.
—Bertrand Russell

Infinity, a concept as vast as the universe itself, has always sparked intrigue and curiosity in human minds. It's the unreachable horizon, the eternal time, the limitless space. It's an idea so large that it's beyond our everyday comprehension. Yet when we delve into the world of fractals, we find ourselves holding infinity in the palms of our hands.

Fractals, with their infinite complexity and self-similarity at all levels, bring the abstract concept of infinity down to a perceptible scale. But how is it possible to handle infinity, let alone see it, touch it, or comprehend it through a simple shape or pattern? This is the paradox of fractals.

From the ancient Greek philosophers to modern mathematicians, the concept of infinity has always been a subject of fascination and mystery. It's an idea that challenges our understanding of the universe and our place within it. Infinity is at once everywhere and nowhere, tangible and intangible, minute and gargantuan. It's a concept that transcends the boundaries of our understanding, yet it's also at the heart of mathematics and the natural world. Fractals, in their infinite complexity, embody this paradox.

The Sierpiński triangle, the Koch snowflake, the Mandelbrot set, and countless other fractal patterns, reveal an infinite complexity within their confines. You can zoom in or out, and the patterns repeat, seemingly forever. Each tiny part contains a reflection of the whole, and the whole is contained within each part. Fractals are both finite and infinite at the same time.

When you delve deeper into the intricacies of these patterns, you are essentially peering into infinity. You see the endless repetition, the unceasing variation on a theme, the constant dance of complexity and simplicity. It's like holding a mirror up to the universe and seeing its infinite reflection looking back at you.

It's not just a mathematical curiosity, but a philosophical one as well. The existence of fractals forces us to reconsider our notions of space, time, and reality itself. They raise questions about the nature of infinity, the structure of the universe, and the limits of our own perception.

How do we make sense of something that is simultaneously finite and infinite? How do we reconcile the vastness of infinity with the tangible reality of a geometric shape? These are not just mathematical questions, but philosophical ones as well. As we delve deeper into the world of fractals, we find ourselves grappling with some of the most profound questions of existence.

The paradox of fractals is a mirror to the paradox of existence. Just as fractals contain infinite complexity within a finite form, so too does our universe contain infinite possibilities within its vast expanse. We are part of that infinity, small yet significant, finite yet interconnected with the infinite cosmos.

Through fractals we see that infinity is not just a theoretical concept, but a tangible reality. It's a reality that we can see, touch, and explore. It's a reality that challenges us, puzzles us, and ultimately expands our understanding of the universe and our place within it. The infinite complexity of fractals is not just a mathematical oddity, but a profound philosophical truth, a testament to the limitless potential of nature, and a reminder of our own infinite possibilities.

The Ancient Greek Perception of Infinity: Pre-Socratics, Socratics, and Zeno

The ancient Greeks, like many ancient civilizations, had an intimate relationship with the concept of infinity. Their approach, however, was unique in its blending of philosophical and mathematical perspectives. The concept of infinity, for them, was as much a metaphysical question as it was a mathematical one.

The pre-Socratic philosophers were among the first to ponder the nature of infinity. Anaximander, for instance, proposed the idea of apeiron, or the boundless. This metaphysical concept was not just about endlessness in space or time, but an underlying principle of the chaotic and indefinite from which all things emerge and return. It's a sort of primordial chaos, formless and infinite, that governs the cosmos.

Following Anaximander, the Pythagoreans, a philosophical and mathematical cult led by Pythagoras, approached infinity through the lens of numbers. They explored the endless series of natural numbers and made profound discoveries in mathematics. Yet the concept of actual infinity, as we understand it today, was unsettling to them. The Pythagoreans revered finite numbers and exact ratios, viewing them as expressions of harmony and order. Infinity, by contrast, represented disorder and chaos. This philosophical discomfort led them to distinguish between potential infinity (an endless process) and actual infinity (a completed totality).

Plato, too, had a complicated relationship with infinity.

He was profoundly influenced by the Pythagoreans and shared their unease with actual infinity. In the *Philebus* Plato argued that the infinite lacks measure and is associated with the chaotic and disorderly. He saw a sort of "negative" infinity that needed to be limited and ordered by the "positive" finite to produce a harmonious cosmos.

On the other hand, Plato's concept of apeiron, akin to Anaximander's "boundless," played a crucial role in his metaphysics. In the *Timaeus* the world soul, which orders the cosmos, is made from the mixture of the Same, the Different, and Being—all of which are derived from the limitless or the infinite.

Zeno of Elea, a contemporary of Socrates, took the paradox of infinity to a whole new level. His famous paradoxes were intellectual puzzles that highlighted the contradictions inherent in our understanding of infinity. The most famous of these is Achilles and the tortoise, where Achilles, despite being faster, can never overtake the tortoise if the latter has a head start. Every time Achilles reaches the point where the tortoise was, the tortoise moves a bit further. Zeno argued that this creates an infinite series of tasks Achilles must accomplish, making it impossible for him to overtake the tortoise.

In essence, Zeno's paradoxes seemed to demonstrate that motion, change, and plurality are illusions because they lead to logical contradictions when we assume that space and time are infinitely divisible. His paradoxes profoundly influenced later philosophical and mathematical treatments of infinity.

The ancient Greek exploration of infinity, as unsettling as it was, laid the groundwork for later developments in mathematics and philosophy. Their metaphysical speculations, mathematical discoveries, and paradoxical puzzles enriched the concept of infinity and set the stage for the remarkable journey to come, a journey that would eventually lead to the fractal geometries we explore today. The infinity that the Greeks grappled with is the same infinity that we grapple with when we delve into the heart of a fractal. It's a testament to the enduring power of their thought that these ancient ideas continue to resonate in the age of fractal geometry.

Anaximander and the Boundless Apeiron

Anaximander of Miletus, a student of Thales and one of the earliest pre-Socratic philosophers, was deeply intrigued by the question of the infinite. He introduced a foundational concept known as apeiron, often translated as "boundless" or "infinite." Anaximander posited this apeiron as a sort of primordial chaos or indefinite matter from which all things emerge and to which all return.

Unlike his mentor Thales, who proposed water as the primary substance of everything, Anaximander argued for something more abstract and indefinable. The apeiron was not merely endless in space or time, but it was also indefinite and indeterminate. It was the limitless from which limits arise—a resource for the generation of all natural things.

In his view, the world we know is only a small orderly part that has emerged from the apeiron. The familiar things we see around us—the earth, the stars, and all living beings—originate from this infinite reservoir and eventually dissolve back into it. Anaximander's apeiron is the eternal source and ultimate end of all things.

This principle of the boundless also served a cosmological role. Anaximander used it to explain the balance and tension between opposites in the world. The hot and the cold, the wet and the dry, all emerge from the apeiron and return to it, maintaining a justice or balance. If one opposite were to prevail, the cosmos would no longer exist.

Anaximander's ideas about the apeiron were profoundly innovative for his time. They marked one of the earliest Greek attempts to grapple with the concept of the infinite, not as a divine being or a physical element, but as a metaphysical principle underlying reality.

In many ways Anaximander's apeiron resonates with the concept of fractals. Both are infinitely complex, both generate a wide array of forms from a simple base, and both blend the finite and the infinite in a dynamic process. Just as Anaximander's cosmos emerges from the interplay of opposites within the boundless, so do fractals unfold from the interplay of mathematical rules within the realm of infinite complexity.

Socrates and the Infinite: A Platonic and Xenophonic Perspective
Socrates, the enigmatic philosopher of ancient Athens, left no written works. Instead, our understanding of his philosophical views comes largely from two of his students: Plato and Xenophon. Through their writings, we glimpse Socrates's thoughts on a myriad of subjects, including his intriguing approach to the concept of infinity.

In Plato's dialogues, Socrates often speaks about the "Forms" or "Ideas," immaterial, perfect entities that exist in an eternal and unchangeable realm. While Socrates does not explicitly discuss the concept of infinity in this context, the idea of the Forms suggests an infinite reality. Each Form is a perfect, timeless archetype of a concept or object in our world, and this implies an infinite realm beyond our material reality. For instance, the Form of Beauty is infinitely beautiful, beyond any beauty we encounter in our tangible world.

In the *Phaedrus* dialogue, Socrates discusses the soul's journey through the cosmos, implying an infinity of experience and knowledge. The soul, he posits, is eternal and imperishable, cyclically incarnating into physical bodies. It has the potential to remember all knowledge it has acquired over infinite time, a concept known as anamnesis.

Xenophon's portrayal of Socrates, while less metaphysical than Plato's, also touches on notions of infinity. In his *Memorabilia*, Xenophon recounts Socrates arguing for the existence of an intelligent and providential divine force. This force, as described, seems capable of perceiving and managing infinite complexity in the cosmos, a task beyond human capacity.

The conception of the infinite in Socratic philosophy, as

depicted by Plato and Xenophon, is subtle and implicit. It's not infinity in a quantitative sense, but rather a qualitative infinity—an infinity of perfection, knowledge, and divine providence.

Once again, we can draw intriguing parallels with fractals. The realm of Forms, with its perfect and infinite entities, resonates with the mathematical infinity within fractals. Fractals, too, embody infinite complexity within simple rules. They can be zoomed in or out indefinitely, always revealing more detail, much like the inexhaustible knowledge Socrates believed our souls could access. Furthermore, the divine intelligence Socrates speaks of in Xenophon's account might be likened to the inherent "intelligence" of the mathematical laws that generate fractal patterns.

Plato and the Infinite: The World of Forms and the Unlimited
Plato's philosophical views, as transmitted through his dialogues, offer a rich tapestry of ideas about the infinite. His notions of the Forms, the "demiurge," and the dichotomy of "the Limited" and "the Unlimited" all hint at a profound understanding of the infinite, though not explicitly in the mathematical sense we associate with it today.

Plato's theory of Forms postulates a realm of perfect, eternal, and unchanging entities, separate from our tangible world. This concept implies an infinity of perfection and immutability. These Forms, whether of Beauty, Justice, or any other abstract concept, exist in their most perfect state, untouched by the temporal and spatial constraints of our physical world. Their realm, though not infinite in number, is infinite in the depth and purity of its realities. Moreover, in the *Timaeus*, Plato introduces the demiurge, the divine craftsman responsible for the creation of the cosmos. The demiurge shapes the cosmos based on the perfect models of the Forms, working with the raw materials of Chaos. This Chaos, characterized by disorder and randomness, represents the infinite in its most undifferentiated state, the Unlimited in contrast to the Limited of the Forms.

In creating the cosmos, the demiurge imposes order on the Unlimited, generating a structured cosmos that mirrors the perfection of the Forms. This cosmic order, while finite, reflects the infinity of the Forms in its perfect harmony and balance.

Just like the demiurge, mathematicians use simple rules to shape the infinity inherent in fractals. Starting from basic shapes (akin to the Chaos), they apply recursive processes to generate infinitely complex patterns. These fractal patterns, while contained within finite spaces, reflect an infinity of detail and self-similarity, much like the cosmos mirrors the infinite perfection of the Forms in Plato's thought.

Plato's philosophy, in its contemplation of perfect Forms and a harmoniously ordered cosmos, captures the spirit of fractals. It reminds us that infinity can be found not only in the vast expanse of the universe but also in the minuscule, self-repeating intricacies of a fractal pattern.

Aristotle and the Potential Infinite

Aristotle, a student of Plato, also grappled with the concept of the infinite. However, his approach was more grounded in the practicalities of the physical world and differed significantly from his teacher's metaphysical abstractions. Aristotle's philosophy is particularly relevant when considering the infinite nature of fractals and their existence in our finite world.

In his work *Physics*, Aristotle rejected the concept of the actual infinite in the physical world. He argued that infinity does not exist in reality, but only as a potentiality. This potential infinite, according to Aristotle, is never realized in its totality but can always be extended. It is not a finished quantity but rather a process that can indefinitely continue.

This understanding of the infinite can be seen as an early intuition of the mathematical concept of a limit, a fundamental tool in calculus, which is itself crucial in the study of fractals. Just as a limit approaches but never reaches a certain value, so too does Aristotle's potential infinite extend without ever becoming a complete, actual infinite.

Aristotle's potential infinite finds a perfect analogy in fractals. When we zoom into a fractal, we find that the complexity and detail can be infinitely extended. However, this infinity is never fully realized within the confines of the fractal's boundary. It remains a potentiality, always capable of revealing more detail but never exhausting its possibilities.

This approach to the infinite allows us to reconcile the paradox of an infinite fractal existing within a finite space. The infinite detail of a fractal is potential, not actual. We can always zoom in further, uncover more detail, but we will never reach an end to this process. Much like Aristotle's potential infinite, the infinity of a fractal is a process, not a completed entity.

By contemplating the philosophies of these ancient thinkers, we gain a richer understanding of the paradoxes inherent in fractals. Their ideas, though formulated centuries ago, provide us with valuable insights into the nature of infinity and its manifestation in the world of fractals.

Zeno's Paradoxes and the Infinite Divisibility

The philosopher Zeno of Elea is another ancient thinker whose ideas echo intriguingly in our modern understanding of fractals. Zeno, a student of Parmenides, is best known for his paradoxes, which illustrate the seemingly absurd implications of infinity. Two of these paradoxes—Achilles and the tortoise, and the dichotomy—have particular relevance for fractals.

In the paradox of Achilles and the tortoise, the fleet-footed Achilles gives a slower tortoise a head start in a race. Zeno argued that Achilles could never overtake the tortoise because, for every distance that Achilles covers, the tortoise moves a little further. The implication is that space can be infinitely divided and that movement across it involves covering an infinite number of these divisions.

The dichotomy paradox further explores this concept of infinite divisibility. It posits that to reach any destination, one must first cover half the distance, then half the remaining distance, and so on. This process results in an infinite number of steps, suggesting that movement is impossible.

At first glance, these paradoxes seem to defy logic. However, they anticipate a crucial concept in the mathematics of fractals: the idea of self-similarity across scales. Much like the infinite divisions of space in Zeno's paradoxes, a fractal pattern repeats at increasingly smaller scales. Each "step" toward the infinite detail of a fractal involves zooming into a smaller scale, much like Achilles's never-ending pursuit of the tortoise or the infinite halving of the dichotomy.

Zeno's paradoxes also underscore the challenge of conceptualizing infinity. The idea of covering an infinite number of divisions or steps is counterintuitive, yet this is precisely what happens when we delve into the infinite complexity of a fractal.

By engaging with Zeno's paradoxes, we gain not only a greater appreciation for the infinite nature of fractals but also a sense of the philosophical puzzles that they embody. The paradoxes of infinity are not just abstract conundrums but fundamental questions about the nature of space, motion, and reality itself—questions that fractals bring vividly to life.

Aristotle and the Potential Infinite

While many pre-Socratic philosophers grappled with the concept of the infinite, it was Aristotle who provided a more nuanced understanding that bears directly on our understanding of fractals. Aristotle differentiated between two types of infinities: the "actual" and the "potential."

The actual infinite is a complete, fully realized infinity. For Aristotle, this concept was illogical, and he rejected it outright. The potential infinite, however, was a different matter. This is an infinity that can always be extended but never fully realized or completed. It's an infinity in progress, so to speak.

This distinction is crucial in the context of fractals. Fractals, in their infinitely repeating patterns, are not examples of an actual infinite. We cannot zoom into a fractal pattern indefinitely in reality because we would eventually reach the limits of the physical universe, such as the Planck length, beyond which current theories of physics do not apply.

Instead, fractals embody the potential infinite. We can always zoom in further and uncover more detail, even though we never reach a final, smallest scale. Similarly, the iterations that generate a fractal pattern can continue indefinitely, even though each individual iteration is a finite step.

Aristotle's potential infinite also resonates with the iterative processes used to create fractals. Each iteration is a finite step toward an end result that is never fully achieved but is continually approximated.

The concept of the potential infinite invites us to see fractals not as static shapes but as ongoing processes, dynamic and ever unfolding. This perspective aligns with a view of the universe not as a fixed and unchanging entity, but as a continually evolving system, filled with patterns that echo across scales and dimensions—a truly fractal cosmos.

Looking back, we can see how these ancient philosophical concepts of infinity—from Anaximander's infinite apeiron, through the Platonic Forms, to Zeno's paradoxes and Aristotle's potential infinite—all resonate with our modern understanding of fractals. These ancient thinkers, in their different ways, were grappling with the same concepts of scale, detail, and infinity that are at the heart of fractal geometry. Their insights, developed in a very different context, still have much to teach us about the nature of fractals and the infinite complexity they embody.

What Fractals Teach Us About Our Understanding of the Universe
While the study of fractals is still relatively young, its potential to reshape our understanding of the universe is immense. From the unfathomable expanses of the cosmos to the microscopic world of atoms and molecules, fractals offer a lens through which we can perceive and comprehend the profound complexity of existence.

For centuries scientists and philosophers have grappled with the nature of reality, each new age bringing with it a paradigm shift that upends what we thought we knew. With the advent of fractal geometry, we find ourselves once again on the cusp of such a shift. Fractals present a new way of understanding not only the structure of the universe but also the processes that shape it.

At the core of this new perspective is the notion of self-similarity, the defining property of fractals. This refers to the repetition of patterns at different scales. In the fractal universe, small parts reflect the whole in a never-ending dance of similarity across scales. This concept has profound implications for how we understand the structure and evolution of the universe.

For instance, consider the branch of a tree, a river network, or the intricate design of our lungs. These natural phenomena exhibit a fractal-like structure, with each part echoing the whole. This concept extends beyond the physical world. Our societies, economies, and even our thoughts and behaviors exhibit fractal-like patterns. The concept of self-similarity suggests that similar processes underlie the formation and evolution of these diverse systems.

Moreover, the infinite complexity inherent in fractals challenges our traditional understanding of dimensionality. Traditionally, we think of the universe as existing within three spatial dimensions. However, the fractal universe is not confined to whole-number dimensions. It can be fractional, further blurring the lines between the large and the small, the one and the many, the finite and the infinite.

This shift in perspective is not just theoretical. It has practical implications for a wide range of fields, from physics and cosmology to biology and computer science. Fractal geometry provides a new language for describing the complexity of the universe. It allows us to model and analyze systems that were previously beyond our grasp, from the structure of galaxies to the dynamics of turbulent fluids, the distribution of species in an ecosystem, or the behavior of financial markets.

The understanding of fractals also challenges our perception of time. In the fractal universe, the future is not entirely unpredictable, nor is the past irretrievable. Instead, they echo each other, each moment in time reflecting a pattern that extends across scales.

Fractals invite us to see the universe as an ongoing dialogue between order and chaos, pattern and randomness, finitude and infinity. They suggest that our existence is woven into the very fabric of this fractal universe, each of us an echo of the whole, a pattern repeating across scales and dimensions.

In this sense, fractals offer a profound lesson about our place in the universe. They remind us that we are not separate from the world around us, but deeply interconnected parts of an intricate and infinitely complex whole. We are, in a very real sense, the universe observing itself through the lens of fractal geometry.

In essence, fractals do more than just present a new way of understanding the universe. They invite us to participate in its unfolding, to marvel at its complexity, and to find our place within its infinite patterns. They challenge us to see the world with new eyes, to question our assumptions, and to appreciate the profound beauty and complexity of the universe we inhabit. The study of fractals, then, is not just a scientific endeavor; it's a philosophical journey, a radical reimagining of our place in the cosmos.

Expanding on the practical implications of fractals, we start to see a pattern of interconnectedness that was previously hidden. Fractal geometry has given us tools to explore and understand the underlying patterns that persist in nature and in man-made systems alike. It's not just about the beauty of the patterns themselves but about how they help us perceive and navigate the world.

In understanding the natural world, fractals have played an indispensable role. For example, the fractal nature of coastlines, a concept explored by Mandelbrot, has helped us better comprehend geographical features and processes. The intricate patterns of veins in leaves, the distribution of galaxies, the pattern of lightning strikes, and the formation of mountain ranges all fall within the realm of fractal geometry. This understanding has far-reaching implications for fields as diverse as geology, meteorology, and environmental science.

In the world of technology, fractal geometry has enabled advancements in digital imaging, graphics, and data compression. It has even influenced the design of antennas and circuits. Recognizing the fractal patterns in these fields has allowed for significant advancements, leading to more efficient technologies and better performance.

In the realm of economics and social sciences, the discovery of fractal patterns in data has led to the development of more accurate models for predicting market behaviors and analyzing social phenomena. This has allowed for more effective decision making and planning.

Fractals also make us reconsider our perception of randomness and order. In the fractal universe, what appears random at first glance often hides a pattern that becomes apparent only when viewed across different scales. This understanding can inform how we approach complex problems, allowing us to find order in what seems like chaos.

In this sense, fractals are not just about mathematics and geometry. They touch on the fundamental questions that have occupied philosophers and scientists for millennia: what is the nature of the universe? How do order and chaos coexist? What is our place in this intricate web of patterns?

By revealing the intricate and often beautiful patterns that underlie the world, fractals challenge us to rethink our assumptions and to seek a deeper understanding. They encourage us to embrace complexity, to be comfortable with uncertainty, and to celebrate the beauty of the universe in all its intricate detail.

In this light, fractals can be seen as a bridge between the physical and the metaphysical, the tangible and the intangible. They invite us to question, to explore, and to marvel at the wonders of the universe. They remind us that we are part of an intricate and infinitely complex whole, and they challenge us to find our place within it.

So as we delve deeper into the world of fractals, we are not just learning about a new kind of geometry. We are embarking on a journey of discovery that touches on every aspect of our lives, from the most fundamental questions about the universe to the most practical applications in science and technology. We are exploring a universe that is at once familiar and profoundly strange, a universe that is infinitely complex yet exquisitely simple, a universe that is, in essence, a fractal.

In the following chapters, we will continue this journey, delving deeper into the implications of fractal geometry for our understanding of the world and ourselves. We will explore how fractals illuminate the interconnectedness of all things, challenge our assumptions about reality, and reveal the profound beauty of the universe. We invite you to join us on this journey, to open your mind to new possibilities, and to see the world through the lens of fractals.

As we venture further into the world of fractals, it becomes evident that they offer more than just an intriguing pattern or an interesting mathematical concept. Instead, they provide a fresh lens through which we can interpret the world around us, a new way to understand reality itself.

At the heart of this new perspective is the realization that complexity and simplicity are not mutually exclusive. Traditional Euclidean geometry, with its straight lines, perfect circles, and uniform planes, tends to convey a sense of simplicity and predictability. The real world, however, is rarely so straightforward. Nature abounds with irregular shapes, uneven surfaces, and complex patterns.

For centuries, this discrepancy posed a significant challenge. How could we reconcile the simplicity of our mathematical models with the complexity of the natural world? Fractals offer a compelling answer. They show us that complexity can arise from simple rules, that simple shapes can generate intricate patterns. They bridge the gap between the neat, predictable world of Euclidean geometry and the messy, unpredictable world of nature.

Through this lens, fractals also challenge our understanding of dimensions. In the Euclidean framework, dimensions are fixed, absolute entities. A line is one-dimensional, a plane is two-dimensional, a cube is three-dimensional, and so forth. But fractals introduce the notion of fractional dimensions, blurring the clear-cut boundaries of Euclidean geometry. In doing so, they invite us to reconsider our assumptions and to view dimensions in a more flexible, nuanced manner.

This new perspective extends beyond the realm of mathematics and into the fabric of reality itself. It encourages us to embrace complexity and to see beauty in irregularity. It invites us to look for patterns in the seemingly random, to seek unity in diversity, to find order in chaos. It shows us that the world is not as predictable, not as straightforward, as we might have thought, but that it is infinitely more interesting and beautiful.

Moreover, the fractal perspective has profound implications for our understanding of time and space, of cause and effect, of order and chaos. It suggests that these concepts too are not as fixed and absolute as we might have thought. Instead, they are fluid, interrelated, and subject to the same principles of self-similarity and scaling that govern fractals.

By challenging our preconceived notions and encouraging us to see the world in a new light, fractals ultimately prompt us to question the very nature of reality. They remind us that the universe is not a collection of discrete, unrelated entities, but a complex, interconnected web of patterns and processes. They hint at a deeper, more holistic understanding of the world, one that recognizes the inherent complexity and interdependence of all phenomena.

In the following sections, we will delve deeper into these ideas, exploring how the fractal perspective can enrich our understanding of various fields, from physics and biology to art and philosophy. Through this exploration, we hope to demonstrate that fractals are not just an intriguing mathematical concept, but a powerful tool for understanding the world, a key to unlocking the mysteries of the universe, a new lens to understand reality.

The concept of fractals, with their infinite complexity, recursive self-similarity, and capacity to span dimensions, has profound implications for our understanding of space and time. It challenges many established notions and brings to the fore new theories that might provide a more accurate representation of the universe.

Let's begin with the concept of space. In classical Euclidean geometry, space is often visualized as a vast, three-dimensional canvas, filled with objects of definite shapes and sizes. This mathematical model of space is clean, tidy, and intuitive. It's also simple, which makes it appealing for basic scientific and engineering applications. However, the regularity of Euclidean space is at odds with the irregularity and complexity of natural phenomena.

Fractals, however, can represent spaces that are not whole numbers—something unimaginable in the Euclidean view of the universe. A coastline, for instance, is not a one-dimensional line nor a two-dimensional surface. It's something in between. It's fractional, jagged, complex. The introduction of fractal geometry by Benoît Mandelbrot in the twentieth century was a revolutionary moment that allowed us to better understand and model natural phenomena.

Now let's turn to time. The conventional understanding of time is linear and unidirectional. It moves forward, never backward, from the past, through the present, and to the future. This is the time of our everyday experience, the time of physics and philosophy. However, the recursive nature of fractals suggests a different perspective.

Fractals allow for scaling, where the same patterns repeat at different scales. Apply this principle to time, and we can visualize history as a fractal pattern, where events of different scales—a day, a year, a century—have similar structures. This challenges the traditional linear view of time and suggests that our perception of time could be much more complex than we imagined.

In the world of theoretical physics, fractals have also made their mark. Scientists have suggested that the fabric of the universe itself may have a fractal structure at the smallest scales, as postulated in theories of quantum gravity. It's an area of ongoing research and, while not proven, the idea that the fundamental structure of the universe might be fractal is a significant shift from established theories.

Furthermore, fractals raise philosophical questions about determinism and free will. Fractals are deterministic: a simple rule generates an infinitely complex pattern. However, the exact form of a fractal can be highly sensitive to initial conditions, a characteristic associated with chaos theory. This paradox, deterministic rules leading to unpredictable outcomes, has profound implications for philosophical debates about determinism and free will.

Fractals, with their inherent complexity and recursive self-similarity, challenge our conventional understanding of space and time. They offer a more nuanced perspective, suggesting that the universe might be more irregular and complex than our traditional models suggest. Whether in the granularity of space or the nature of time, fractals force us to reconsider our assumptions and provide a fresh perspective on longstanding philosophical and scientific debates.

The concept of fractals stands in stark contrast to the principles of Euclidean geometry, which has been the bedrock of our understanding of space for over two millennia. Euclidean geometry is clean and orderly, dealing with shapes and figures that have whole number dimensions—lines (one-dimensional), squares (two-dimensional), and cubes (three-dimensional), to name a few. It is intuitive and aligns neatly with our day-to-day experiences.

But when we try to apply Euclidean principles to the natural world, we often find that they fall short. Take, for example, the seemingly simple task of measuring a coastline. If you measure it with a mile-long ruler, you will get a certain length. However, if you measure it with a foot-long ruler, the total length increases because the smaller ruler can fit into smaller nooks and crannies. This phenomenon, known as the coastline paradox, was a confounding problem until the introduction of fractal geometry.

Fractals brought a new perspective. Unlike Euclidean shapes, fractals can have dimensions that are not whole numbers, allowing them to better represent the intricate complexity of natural phenomena. The dimension of a fractal is a measure of how much space it fills as you zoom in on it, allowing it to capture the detail of a coastline or the intricacy of a snowflake at any scale.

Moreover, the recursive nature of fractals, where a simple pattern repeats infinitely at varying scales, challenges the Euclidean principle of self-identity. In Euclidean geometry, a shape remains the same regardless of how you scale it. A square remains a square whether it's two inches wide or two miles. But with fractals, the shape changes as you zoom in or out, revealing new levels of complexity.

The shift from Euclidean geometry to fractal geometry in describing the world around us has significant implications. It not only offers a more accurate representation of the natural world but also challenges our fundamental understanding of space. It asks us to reconsider our notions of order, complexity, and dimensionality, reshaping our perceptions of the world in the process.

Traditionally, our understanding of time has been linear and unidirectional. We think of time as a straight line, a continuous sequence of moments flowing from the past, through the present, and into the future. This perspective is deeply ingrained in our consciousness and language—we speak of moving forward into the future and leaving the past behind.

However, the concept of fractals offers a different perspective that challenges this conventional understanding. If we think of time as a fractal dimension, then time isn't merely a line but a complex structure, filled with loops, twists, and turns.

In a fractal conception of time, the past, present, and future are deeply interconnected, much like the repeating patterns in a fractal. Events echo across time, repeating and reverberating in complex patterns that we're only beginning to understand. Just as a fractal shape contains self-similar patterns at every level of zoom, the fractal nature of time suggests that patterns repeat at different scales—hours, days, years, centuries, and so on.

To visualize this, imagine looking at a fractal pattern. As you zoom in, you would see the same pattern repeating over and over, at smaller and smaller scales. Now apply this to time. Rather than being a straight line, time becomes a tapestry of interwoven patterns, with similar events and phenomena recurring on different scales.

This perspective does more than just challenge our linear concept of time; it also has profound implications for our understanding of causality, history, and destiny. It suggests that the past and the future are deeply intertwined, and that events may echo across time in ways that are predictable, if only we understood the patterns.

However, it's important to note that this perspective is speculative and metaphorical. While it provides a fascinating way to think about time, it doesn't replace our conventional understanding. Time, like all concepts in physics, is ultimately a tool that we use to understand and navigate the world. Whether we think of it as a line or a fractal, the true nature of time is likely to remain a mystery.

Fractals and the Challenge to Causality

Causality is a concept deeply rooted in our everyday experience and scientific understanding of the universe. It's the principle that everything that happens (an effect) is due to some prior event (a cause). It's the reason why, when we drop a glass, we expect it to fall and shatter, and not suddenly float upward or transform into a butterfly. Our world, as we understand it, is fundamentally causal.

However, the fractal nature of time, as previously discussed, presents us with a different perspective that challenges this causal understanding of the universe. If time is indeed fractal—if it loops, twists, and turns, with patterns repeating at different scales—then the linear, one-way flow of cause to effect could be called into question.

In the fractal perspective, events aren't merely linked in a simple cause–effect chain, but are part of an intricate, self-similar pattern that extends across time. An event in the present might echo a similar event in the past and resonate with another in the future. In this view, the future can influence the past just as the past influences the future.

This notion seems to contradict our everyday experience and the laws of physics as we currently understand them. However, it's worth noting that at the quantum level, the line between cause and effect can become blurred. For instance, in certain quantum experiments, the choice of measurement can appear to retroactively determine the state of a particle in the past, a phenomenon known as quantum retrocausality.

The fractal conception of time and its implications for causality remain speculative and largely metaphorical. They are not currently supported by empirical evidence or widely accepted in the scientific community. However, they offer a fascinating perspective and a challenge to our conventional understanding of time and causality.

It's possible that these ideas may inspire new ways of thinking about the nature of time, causality, and the universe itself. As with all radical ideas, they should be approached with an open mind, a healthy dose of skepticism, and a willingness to follow where the evidence leads.

The concept of four-dimensional space is challenging to grasp because we're accustomed to living in a three-dimensional

CHASING DRAGONS BETWEEN DIMENSIONS

world. However, let's consider a hypothetical 4D shape, known as a hypercube or tesseract, to visualize the concept.

Imagine a point in space. This is a zero-dimensional object. If we extend this point in a direction, we get a line, a one-dimensional object. If we take this line and extend it perpendicular to itself, we get a square, a two-dimensional object. Further, if we extend the square perpendicular to the plane it lies in, we get a cube, a three-dimensional object.

Now, if we extend this cube in a direction perpendicular to the three dimensions we're familiar with, we'd have a tesseract, a four-dimensional object. A tesseract is to a cube what a cube is to a square. The tricky part is visualizing this, as we can't directly perceive the fourth spatial dimension.

However, we can represent a tesseract in three dimensions, just like we can represent a cube on a two-dimensional piece of paper. When we do this, the tesseract looks like a cube within a cube, with the corners connected by lines (these lines represent the fourth dimension).

In the context of the universe and the concept of time, the tesseract metaphor can be applied. If we imagine our universe as a 3D slice of a 4D hypercube, each moment in time could be seen as another slice. Our experience of time could then be thought of as moving through these slices, one after another.

Now let's consider the implications of fractals and the challenge to causality within this 4D framework. If time is indeed fractal, then the pattern of events is not linear but self-similar across different scales. Within the tesseract, this could be visualized as intricate, repeating patterns of connections stretching across the different slices of time.

In terms of causality, if the future can influence the past as suggested by the fractal conception of time, it would be as if connections within the tesseract could run backward as well as forward. This is a mind-bending concept that challenges our usual understanding of cause and effect.

Of course, this is all highly speculative and metaphorical. It should be taken as a thought experiment, a way of visualizing concepts that are beyond our usual experience, rather than a literal description of reality.

Continuing from the fractal conception of time and its implications, let's now delve into the realm of quantum mechanics, where the strange becomes even stranger. Here, the fractal structure of time could tie into the notion of quantum superposition and entanglement.

In quantum mechanics, particles can exist in a state of superposition, meaning they can be in multiple states at once, only collapsing to a single state when observed. Now consider our earlier tesseract metaphor: if time is indeed a higher-dimensional fractal, could this mean that events, like quantum particles, are in a superposition of states until observed? This idea may sound far-fetched, but it does resonate with some interpretations of quantum mechanics.

Moreover, quantum entanglement, the phenomenon where particles become interconnected such that the state of one instantaneously affects the state of another, regardless of the distance between them, presents an even more intriguing twist. If we imagine these entangled particles existing in our fractal tesseract, it seems as though they are somehow bypassing the usual flow of time. This apparent "action at a distance" fits oddly well with the notion of a fractal universe, where patterns repeat across scales and seemingly distant points can be intrinsically connected.

This line of thinking opens a Pandora's box of questions. If we live in a fractal universe, does it imply that reality itself is not as fixed as it seems? Could the future, present, and past exist simultaneously, with time being nothing more than the path we take through these moments? And if so, what does that mean for our understanding of cause and effect, free will, and the nature of consciousness itself?

The fractal concept is truly a tool for challenging and expanding our perspectives on the universe. While we don't have all the answers, and many of these ideas remain in the realm of speculation, they represent a radical departure from our everyday perception and provide fertile ground for thought and exploration.

Just as the exploration of fractals has broadened our understanding of dimensions, the application of fractal concepts to time and the universe may lead us to novel insights and, perhaps, a better understanding of the very nature of reality. As we continue to delve deeper into the world of fractals, we may find that these intricate patterns provide not just a window into the mathematical elegance of the universe, but also a mirror reflecting our own understanding and experience of reality. And that, in itself, is a fascinating journey worth embarking on.

As we delve into the realm of speculation, let's imagine a machine that could navigate the fractal dimensions between three and four, potentially enabling time travel. To understand how this might work, we must first understand the concept of fractal dimensions and the idea of time as a dimension itself.

The fractal dimensions are those that lie between the whole number dimensions we're familiar with. Just as the coastline of Britain, a classic example of a fractal, is more than one-dimensional but not quite two-dimensional, so a fractal dimension might be more than three-dimensional but not quite four-dimensional.

Time, according to the theory of relativity, can be considered as a fourth dimension. Just as we move through three-dimensional space, we also move through time, though in a much more constrained way: always forward, never backward or sideways.

Now, let's imagine a machine, an advanced piece of technology that allows us to explore fractal dimensions. This "fractal navigator" would need to be able to manipulate the very structure of space-time in a way that we currently can't fathom. It would have to stretch, squeeze, and contort our familiar three dimensions, opening pathways into the fractional dimensions that exist between them.

In practice, operating this machine might be somewhat

like using a complex flight simulator. The user would input a set of coordinates, representing a location not only in three-dimensional space but also in fractal-dimensional space. The machine would then compute a trajectory through this higher-dimensional space, allowing the user to "jump" from one point in our three-dimensional universe to another.

If time is indeed a fourth dimension, then moving through fractal dimensions might allow us to move through time in unconventional ways. The fractal navigator might let us take shortcuts through the fractal structure of space-time, effectively traveling to the past or future.

Of course, this is all purely hypothetical. We don't have any concrete evidence that such a machine could be built, or that time travel is even possible. Moreover, the idea of fractal time travel raises a whole host of paradoxes and problems that we don't currently have solutions for.

However, the very act of imagining such a machine is a valuable exercise. It pushes us to think outside the box, to challenge our assumptions, and to explore the fascinating possibilities that fractals open. Who knows what discoveries might be waiting for us as we delve deeper into the fractal universe?

As we reach the conclusion of this chapter, it's clear that fractals offer an entirely new lens to view and understand the universe. They stand at the intersection of mathematics, philosophy, and the very nature of reality, offering insights that challenge and expand our most fundamental assumptions.

In this whirlwind journey, we've seen how fractals break down the barrier between the finite and the infinite, echoing the paradoxes of ancient Greek philosophy. We've explored how they challenge our traditional Euclidean understanding of space, introducing the concept of fractional dimensions and infinitely complex structures arising from simple rules.

Moving from space to time, we delved into the intriguing proposition of fractal time—a concept that not only alters our perception of past, present, and future, but also ties into the strange and counterintuitive phenomena of quantum mechanics. The idea of time as a higher-dimensional fractal aligns uncannily well with concepts like superposition and entanglement, raising questions about the nature of reality, causality, and consciousness.

We've also considered the potential of fractals as a stepping stone toward understanding and visualizing higher dimensions. Just as fractals bridge the gap between dimensions in space, could they also be key to unlocking the mysteries of the fourth dimension and beyond? This line of inquiry opens a world of possibilities—including some that seem to belong more to the realm of science fiction than science.

One such possibility is the concept of time travel. If time is indeed a fractal, with all moments existing simultaneously, it stands to reason that traveling through time might merely involve moving along a different path through the fractal structure of the universe. While this idea is purely speculative and fraught with theoretical challenges—not to mention the paradoxes inherent to time travel—it's a tantalizing thought.

Of course, we are far from developing a "fractal time machine." Still, the very fact that such wild ideas are even on the table speaks volumes about the transformative potential of fractals. They represent a radical shift in perspective, challenging us to rethink everything we thought we knew about the universe.

In the end, the true magic of fractals may lie not in the answers they provide, but in the questions they raise. As we continue to explore these fascinating patterns, who knows what new insights and possibilities might unfold? One thing is clear: when it comes to fractals, we've only just scratched the surface. The journey into the fractal universe is just beginning—and the road ahead is as infinitely complex, surprising, and beautiful as the fractals themselves.

Let's indulge our imagination further and envision using this fractal navigator, this gateway to the infinite, to revisit the ancient thinkers who first pondered the concept of infinity.

Imagine stepping into the fractal navigator and inputting the coordinates for ancient Miletus. As the machine hums to life, reality shifts and bends around you. Then, suddenly, you're standing in the bustling marketplace of the ancient city. In front of you stands Anaximander, deep in thought, gazing at the patterns of nature and the cosmos. You approach him, greeting him in his native tongue, and embark on a discussion about the infinite, about the boundless apeiron, and the fractal nature of reality. You share your understanding of fractals, of dimensions beyond the third, of the very machine that made this conversation possible. You see his eyes light up with fascination and curiosity, his mind grappling with these concepts that are millennia ahead of his time.

Next you input the coordinates for ancient Athens, and in an instant, you're walking the streets where Socrates once roamed. You find him in the Agora, surrounded by his disciples, engaged in his characteristic style of questioning. You join the conversation, introducing the concept of fractals, explaining how they challenge our conventional notions of space and time. Socrates, ever the skeptic, asks probing questions, forcing you to elucidate your ideas further, pushing you to explore the implications of these ideas.

Then it's onto the Academy, where Plato is deep in thought, contemplating the realm of the Forms. You share with him how fractals might represent a bridge between the perfect world of Forms and the imperfect, material world we inhabit. You discuss the Platonic solids, the five regular polyhedra that Plato associated with the elements, and speculate on the possible existence of fractal solids that extend into higher dimensions.

The journey continues, from philosopher to philosopher, each encounter enriching your understanding of infinity, fractals, and the nature of reality itself. As you converse with these ancient thinkers, you realize that the dialogue of philosophy, much like the structure of a fractal, is never-ending. Each question leads to further questions; each answer reveals deeper mysteries; each conversation is a dive into the infinite.

As you return to your own time, stepping out of the fractal navigator, you carry with you not only a deeper understanding of fractals and infinity but also a profound sense of connection to the great philosophical tradition that stretches back to the dawn of Western thought. Through your journey across fractal dimensions, you've reached across the vast expanse of time, engaged in dialogue with the great thinkers of the past, and glimpsed the boundless potential of the human mind to explore the infinite. The words of Socrates echo in your mind: "I know nothing except the fact of my ignorance." You realize that, with fractals, you've uncovered a whole new realm of ignorance—and a whole new realm of potential knowledge. The journey, it seems, is only just beginning.

4: FRACTALS IN NATURE AND CULTURE

It is human nature to want to exchange ideas, and I believe that, at bottom, every artist wants no more than to tell the world what he has to say. I have sometimes heard painters say that they paint "for themselves" but I think they would soon have painted their fill if they were completely alone in the world. So, we…labor to make our thoughts visible, or, if you like, we invent and tell lies to that end. Now, what is it that we want to say? It is a bit difficult to find the right word for it, but I suppose one could describe it as a wish for a system, or, more exactly, the quest for such a system. The system that we seek must be based on an indefinite plurality of repeated elements. For me, it remains essential that each component be identical to the next.
—M. C. Escher

Fractals in the Natural World

The allure of fractals isn't confined to the realm of abstract mathematics or philosophical conjecture; they have a tangible, visible presence in the world around us. The infinite complexity of fractals, it turns out, resonates with the structure and pattern we find in our natural surroundings. By studying fractals, we're not just engaging with mathematical curiosities; we're decoding the language of the universe itself.

Consider, for instance, the branching of trees. Each branch

divides into smaller branches, which divide into even smaller branches, in a pattern that repeats across scales. This self-similarity is a defining characteristic of fractals. The veins in a leaf, too, follow a fractal pattern, spreading out to cover an increasingly large area with each new level of branching. Even if we zoom in on a small section of the leaf, we see the same pattern repeating.

Rivers and coastlines also display fractal characteristics. The course of a river, viewed from above, looks remarkably like a tree branch or a leaf vein. Similarly, coastlines, when viewed from different distances, exhibit the same jaggedness, the same intricate detail. This is known as the coastline paradox, a concept introduced by Benoît Mandelbrot himself: the length of a coastline appears to be infinite because it gets longer the more closely you look at it.

Fractals even reach into the sky. The pattern of lightning as it forks across the sky, the shape of a cloud, the distribution of galaxies across the universe—all these display fractal patterns. The beauty and complexity we see in the night sky, in the landscape around us, in the tiniest leaf or the mightiest river, all echo the beauty and complexity of fractals.

The harmony of fractals with nature extends to life forms as well. The structure of our lungs, for example, is fractal, with bronchi branching off into smaller bronchioles in a repeating pattern that allows for efficient oxygen exchange. The circulatory system, too, is fractal, with large arteries branching into smaller arterioles and then into capillaries. The same pattern is evident in the neurons in our brain. The fractal design allows for maximum efficiency in a compact space.

From the infinitesimal to the infinite, from the terrestrial to the celestial, fractals provide a framework to understand the pattern and structure of the world around us. They are not just mathematical constructs but a language that nature speaks. When we study fractals, we're learning to understand that language, to see the world with new eyes.

Fractals are woven into the fabric of our existence, and by

understanding them, we gain insights into the intricate patterns and beautiful complexity of the universe we inhabit. As we continue our journey into the world of fractals, let us carry this understanding with us, and let it inspire us to look at our world and ourselves in a new light.

Tree and Plant Branching

There's a simple, yet profound beauty in the branching structures of trees and plants. This branching pattern is a prime example of fractals in nature. To understand how, let's start from the very basics. Imagine a seed. This seed is the starting point, the "zeroth" iteration, if you will, of a potentially magnificent tree. As the seed sprouts, it gives way to a stem, which then bifurcates into branches. This is the first iteration. Each of these branches further bifurcates, giving rise to more branches. This is the second iteration. As this process continues, a complex, dendritic structure begins to emerge, characterized by a repeating pattern of bifurcation. This self-similarity across scales is the hallmark of fractal structures.

But what's truly intriguing is the mathematical precision behind these organic forms. The angle at which branches split off, the ratio of branch lengths, the rate of bifurcation all follow specific mathematical rules. For instance, many trees exhibit what's known as the Fibonacci sequence in their branching pattern. Each number in the Fibonacci sequence is the sum of the two preceding numbers, starting from zero and one. In the context of trees, this might represent the number of branches at each level of bifurcation.

The application of such mathematical principles is not a mere coincidence. Rather, it reflects an optimization strategy by nature. Fractal branching allows trees and plants to maximize their exposure to sunlight and nutrients, thus enhancing their chances of survival.

Interestingly, this pattern isn't just limited to the macroscopic structure of trees. It extends down to the microscopic level as well, such as the venation patterns on individual leaves, which we will explore in more detail in the next section.

In sum, the fractal nature of tree and plant branching represents a marvelous interplay between mathematical principles and evolutionary optimization. It serves as a testament to the inherent efficiency and aesthetic beauty of fractals, which permeate every level of our natural world.

River Networks

The branching pattern of river networks is another classic example of fractals in nature. When seen from a bird's-eye view, a river system is revealed to be an intricate network of tributaries that merge to form larger rivers, finally culminating in a main river channel that drains into a sea or ocean. This hierarchy of branching, reminiscent of the dendritic structure of trees, shows a striking self-similarity across different scales.

The formation of these fractal river networks can be traced back to the interplay of several environmental factors like topography, rainfall, and soil composition. However, at the heart of it all lies a simple iterative process. Each raindrop that falls on the landscape contributes to the runoff that carves out tiny channels on the surface. As more rain falls, these channels merge to form larger channels, which eventually become streams, then rivers. This iterative process of erosion and channel formation gives rise to the fractal structure of river networks.

One of the fascinating aspects of this fractal structure is the mathematical relationship known as Hack's law. Proposed by the geologist Luna Leopold and hydrologist John Hack, this law states that the length of a river is proportional to the area of its drainage basin raised to a certain power. This power, which is typically close to 0.6, is a measure of the river network's fractal dimension.

The fractal nature of river networks isn't just a curiosity—it has important implications for a range of fields. For instance, it helps hydrologists predict flood patterns and design more efficient drainage systems. It also provides clues to geologists about the history of a landscape and the forces that have shaped it.

River networks represent yet another example of the fractal geometry that pervades our natural world, a manifestation of the complex interplay of simple rules and environmental factors.

Mountain Ranges

When viewed from afar, the jagged skyline of a mountain range presents a breathtaking sight. But beyond the natural beauty, there's a mathematical wonder at play. Mountain ranges, with their rugged landscapes, exhibit a fractal-like structure.

The fractal nature of mountains is apparent when we consider their irregular shape and the self-similarity they exhibit across different scales. Whether you're looking at an entire mountain range or a single peak, you'll see a similar pattern of ridges, valleys, and peaks.

This self-similarity can be understood better through the concept of the fractal dimension. A perfectly smooth, flat land would have a dimension of two, whereas a completely filled, three-dimensional volume would have a dimension of three. The fractal dimension of mountainous landscapes falls between these two extremes and is typically close to 2.5. This means that mountains occupy more space than a flat plane but less than a full volume, reflecting their irregular, fractal nature.

But how do mountains get their fractal shape? The answer lies in the interplay of geological processes like erosion and plate tectonics. Over millions of years, these processes shape the landscape, carving out valleys and thrusting up peaks. Each of these processes operates in a somewhat random yet iterative manner, much like the recursive algorithms used to generate mathematical fractals.

The recognition of mountains as fractals has had significant implications, particularly in computer graphics and geology. For instance, fractal algorithms form the backbone of many terrain generation software, allowing for the creation of realistic virtual landscapes. Meanwhile, in geology, understanding the fractal nature of mountains can help researchers model erosion processes and predict future landscape changes.

In essence, mountains serve as a testament to the fractal patterns that emerge from the slow, relentless workings of nature's forces. Their jagged peaks and valleys, formed over millions of years, encapsulate the sublime beauty of fractal geometry.

Clouds

We often associate the shape of clouds with randomness and chaos, yet their formation is not random at all. In fact, the shapes of clouds can be understood through the lens of fractal geometry. Like coastlines, mountain ranges, and rivers, clouds too exhibit fractal properties, specifically self-similarity across multiple scales.

When we observe clouds, we see an array of forms: cumulus, cirrus, stratus, and many more. Each type possesses its own characteristic shape and texture, but all share the quality of self-similarity. Whether you're looking at a cloud-filled sky or a single fluffy cumulus, the fluffy, uneven structures are strikingly similar.

The fractal dimension of clouds can be understood through their formation process. Clouds form when humid air rises and cools, causing the water vapor to condense into tiny droplets. This process, known as convection, creates the familiar fluffy structures we see in the sky. The water droplets cluster together, forming larger and larger groups in a random yet structured way, creating a pattern that repeats on different scales—a key characteristic of fractals.

Understanding the fractal nature of clouds has practical applications. For instance, in meteorology, it helps improve weather prediction models. By treating clouds as fractals, meteorologists can simulate their formation and evolution more accurately. In computer graphics, fractal algorithms are used to create realistic clouds in digital environments, contributing to more immersive experiences in video games and films.

So the next time you gaze up at the sky, remember that the clouds you see are more than just random shapes. They're visual manifestations of the intricate fractal patterns that govern many aspects of our natural world.

Plant Growth

The growth of many plants exhibits fractal patterns, from the arrangement of leaves on a stem to the branching of trees and the patterns seen in ferns. This self-similar repetition at increasingly smaller scales is a remarkable feature of nature, evident in a variety of plant forms.

Consider the arrangement of leaves around a stem, a phenomenon known as phyllotaxis. Many plants exhibit what's called Fibonacci phyllotaxis, where the number of turns between successive leaves follows the Fibonacci sequence (1, 1, 2, 3, 5, 8,...). This pattern ensures optimal exposure to sunlight and rain for each leaf. When viewed from above, the arrangement of leaves forms a spiral pattern that repeats on different scales, making it a perfect example of a fractal in nature.

The branching patterns of trees and plants also demonstrate fractal characteristics. A tree's growth can be modeled using a simple recursive algorithm, where each branch splits into smaller branches, which in turn split into even smaller branches, and so on. This self-similar branching pattern is seen across many species, from oak trees to broccoli.

Ferns provide another compelling example. If you look closely at a fern leaf (frond), you'll notice that each small leaflet (pinna) resembles a miniaturized version of the whole frond. This self-similarity continues down to even smaller scales.

Understanding the fractal nature of plant growth has significant implications. It helps botanists and ecologists understand plant development and structure, aids in the development of more accurate models for forest ecology, and even inspires designs in architecture and computer graphics.

In sum, plant growth is a testament to the ubiquity of fractals in the natural world. It shows how a simple rule, repeated over and over, can produce complex and beautiful patterns, much like a fractal.

Seashells

Seashells, the exoskeletons of marine mollusks, exhibit fractal patterns in their shape and structure. The characteristic spiral shape of many shells is a direct result of a simple, repetitive growth process—the essence of fractal geometry.

Take the nautilus shell as an example. The nautilus is a marine creature that builds its shell by adding on new chambers as it grows. Each new chamber is a scaled-up copy of the previous one, resulting in a logarithmic spiral that maintains its shape at all scales—a fractal pattern. This spiral follows the Fibonacci sequence, a series of numbers where each number is the sum of the two preceding ones (0, 1, 1, 2, 3, 5, 8, 13,...), which is intrinsically related to the golden ratio, a mathematical constant found in many natural and man-made structures.

Similarly, the conch shell grows in such a way that each whorl is a scaled version of the previous one, producing a shape that is self-similar across scales. The self-similar growth pattern allows the shell to maintain a consistent shape while accommodating the growing animal within.

Fractal patterns in seashells are not limited to their overall shape. The surface of a shell often exhibits intricate patterns of ridges, spines, or bumps, which are themselves self-similar. These patterns are thought to enhance the shell's strength or camouflage ability.

Studying the fractal patterns in seashells helps scientists understand the growth processes of marine mollusks and the evolutionary advantages of these patterns. For mathematicians and artists, seashells are beautiful real-world examples of how simple rules can generate complex and aesthetically pleasing forms. Just as they do in trees, mountains, and clouds, fractals in seashells remind us of the profound unity underlying nature's diversity.

Corals

Corals are an excellent example of three-dimensional fractals found in nature. These marine invertebrates are known for their colorful and intricate structures, which, upon close examination, reveal a repeating pattern that is a hallmark of fractal geometry.

Corals belong to the phylum Cnidaria, which also includes sea anemones and jellyfish. Many species of corals, especially those in the *Acropora* genus, are characterized by complex branching structures. A typical coral colony is composed of hundreds to thousands of individual polyps, which are tiny, soft-bodied organisms. These polyps secrete a hard limestone skeleton that forms the backbone of the colony. As the polyps continue to reproduce and secrete more skeleton, the coral colony grows and expands, often in a fractal-like pattern.

The branching patterns in corals are self-similar, meaning each smaller branch is a miniature, albeit imperfect, copy of the larger structure. This pattern repeats at different scales, creating a complex, three-dimensional fractal pattern. In this sense, looking at a single branch of coral can give you an idea of the structure of the entire colony.

The fractal nature of coral structures is not merely a visual spectacle. It has significant biological implications. The complex, fractal-like structure of corals increases their surface area, allowing a greater number of polyps to inhabit a given volume of space. This enables the coral colony to capture more sunlight and nutrients, enhancing its survival in nutrient-poor tropical waters.

Furthermore, the fractal geometry of corals contributes to their resilience. The repeating, interconnected structure provides a buffer against damage. If one part of the colony is damaged, the rest of the colony can continue to function because of the redundancy built into its fractal architecture.

The study of fractals in corals also has implications beyond biology. Researchers have used fractal geometry to model the growth of coral reefs, predict their response to environmental changes, and develop strategies for coral conservation.

Corals, with their vibrant colors and intricate fractal patterns, serve as a vivid reminder of the pervasiveness of fractals in our natural world. They illustrate how the abstract mathematics of fractals can manifest in tangible, biological structures, and they underscore the deep connections between geometry, biology, and ecology.

The Example of Sponges

Sponges, like corals, provide another striking illustration of three-dimensional fractals in the natural world. Sponges are marine animals known for their porous structures, which facilitate the flow of water and nutrients through their bodies. The structure of many sponge species, particularly those in the Scleractinia family, exhibit fractal-like patterns.

Sponges are part of the phylum Porifera, known for their unique feeding system. Sponges draw in water through tiny holes or pores on their surfaces. These pores lead to a network of internal canals and chambers where the sponge's cells extract nutrients and oxygen from the water. The water is then expelled through larger openings called oscula. This entire process is driven by the beating of whiplike structures called flagella on the sponge's choanocytes or "collar cells."

The structure of a sponge is specifically designed to facilitate this feeding mechanism. The system of canals and chambers within a sponge is complex and intricate, and in many species it exhibits a fractal-like pattern. This fractal architecture maximizes the sponge's surface-area-to-volume ratio, allowing it to filter a larger volume of water and extract more nutrients.

Like corals, the fractal nature of sponges' structure isn't just visually captivating but also biologically significant. The fractal network of pores and canals is robust and adaptable, allowing the sponge to withstand various environmental pressures. Moreover, the redundancy in the fractal structure means that if one part of the sponge is damaged, the rest of the organism can continue to function.

The study of fractals in sponges extends to various scientific domains. For instance, in material science, the fractal structure of sponges has inspired the design of novel materials with high surface area, flexibility, and resilience. In the field of fluid dynamics, the fractal network of canals in sponges has been used to model how fluids flow through complex networks.

The example of sponges illustrates how fractal geometry, an abstract mathematical concept, finds practical and tangible expression in the natural world. By studying these natural fractals, we gain insights into the principles that guide biological design and the ways in which organisms adapt to their environment. This knowledge, in turn, can inform and inspire human innovation, from material design to fluid dynamics.

Fractal Patterns in Animal Markings

Animal markings, such as the stripes of a zebra or the spots of a leopard, often display fractal-like patterns. While not a perfect mathematical fractal, the patterns do exhibit a degree of self-similarity and complexity that is reminiscent of fractal structures. Let's dive into two specific examples: the coat patterns of a leopard and the shell patterns of a certain type of mollusk known as the cone snail.

Leopards: The spots on a leopard, known as rosettes, create a pattern that bears similarity to certain types of fractals. Each rosette is a complex pattern in itself, with a dark outline surrounding a lighter interior that often contains a smaller spot. This pattern is repeated across the leopard's body, creating a sort of self-similarity. The rosettes are also dispersed in a nonuniform manner across the leopard's body, which contributes to the complexity of the overall pattern. Scientists believe that these complex patterns aid in camouflage, breaking up the leopard's outline in a variety of habitats from grassland to shadowy woodland.

Cone Snails: Cone snails, or Conus, are a large genus of small to large predatory sea snails. The most striking aspect of these creatures is their highly geometric, intricately patterned shells. The patterns on these shells often display a high degree of complexity and self-similarity that is reminiscent of fractal patterns. For example, a pattern might consist of a series of lines or dots that are repeated at smaller scales across the shell. The exact pattern varies between species and individuals, leading to a stunning variety of shell designs. These patterns are thought to play a role in camouflage and species recognition.

In both these cases, while the patterns may not be perfect mathematical fractals, they do exhibit a degree of complexity and self-similarity that is characteristic of fractal structures. It's fascinating to consider how these patterns, created through biological processes, reflect the same principles of geometry that we see in the abstract world of mathematics.

Ant Colonies and Fractals

Ant colonies are another fascinating example of natural phenomena that display fractal-like patterns. While an individual ant might seem simple, the collective behavior of an ant colony can lead to the creation of complex structures and patterns that bear resemblance to fractals.

Nest Construction: One of the most prominent examples of this is in the construction of ant nests. Ant nests are intricate structures with a high degree of complexity. They often consist of a network of tunnels and chambers that extend deep into the ground, exhibiting a self-similar branching pattern that is characteristic of fractals.

This self-similar branching is not a coincidence but is a product of the collective behavior of the ants. Each ant follows a set of simple rules, such as digging in a certain direction or depositing pheromones, which leads to the emergence of complex, large-scale structures. This process, known as stigmergy, is a form of indirect communication through the environment, and it's a key mechanism through which simple creatures like ants can build complex, fractal-like structures.

Foraging Patterns: Another area where we see fractal patterns is in the foraging behavior of ants. When ants are searching for food, they often create trails that exhibit a fractal pattern. This is because the trails are a form of "random walk," where each step is determined by a combination of the ant's individual decisions and the pheromone trails left by other ants. The resulting pattern is a complex, branching network of trails that is self-similar across different scales, much like a fractal.

Again, the fractal-like patterns arise from the collective behavior of the ants. Each ant follows a set of simple rules, but when these rules are repeated across the colony, they lead to the emergence of complex, large-scale patterns.

These examples show how fractal-like patterns can emerge from the collective behavior of simple creatures. Through the simple rules they follow and the interactions between them, ants can create complex structures and patterns that have a high degree of complexity and self-similarity, much like fractals.

Bird Nests and Fractals

Bird nests are yet another example of how fractal patterns appear in the natural world, particularly in the behavior and architecture of animals. While a bird's nest might not seem to be as mathematically precise as other examples of fractals in nature, the process and approach birds use to build their nests reflect the repetitive, iterative process that characterizes fractal generation.

Birds build their nests using a variety of materials, including twigs, leaves, feathers, and even man-made debris. The construction process can be seen as an iterative one, where the bird repeatedly brings materials back to the nest site and arranges them in a particular way. This process of building up layers of material is analogous to the iterative process used in constructing fractals.

While the overall shape of the nest may not be strictly self-similar, the iterative process used in its construction reflects the principle of recursion that is central to fractals. Additionally, when viewed at different scales, the arrangement of materials within the nest can exhibit a degree of self-similarity, particularly in nests with more complex structures.

Beyond the nests themselves, fractal patterns can also be seen in the nesting behavior of some bird species. For example, some birds exhibit territorial behavior where they space their nests in a pattern that maximizes their access to resources while minimizing conflict with neighbors. This often results in a dispersed pattern of nests that shows a degree of self-similarity across different scales, like fractal patterns.

Birds that nest in colonies, such as penguins or flamingos, also show fractal-like patterns in their nesting behavior. The arrangement of nests within the colony often follows a power law distribution, a type of pattern that is associated with fractals. This pattern results from the balance between the benefits of nesting close to others (for protection and social interaction) and the costs (such as competition for resources).

While bird nests and nesting behavior might not exhibit the strict mathematical precision of geometric fractals, they reflect the principle of recursion and self-similarity that characterizes fractals. As such, they offer another intriguing example of how fractal patterns can emerge from the behavior and activities of animals in the natural world.

Mountain Ranges and Coastlines

Few natural formations exemplify fractal geometry as clearly as mountains and coastlines. Both these formations, while vastly different in their nature and formation process, share a common characteristic: self-similarity across different scales, a key feature of fractal geometry.

Mountain ranges, with their rugged and complex topography, provide a compelling case for the presence of fractals in nature. Take any mountain range—the Himalayas, the Rockies, or the Andes, for instance. At a macro level, the entire range exhibits a rough and irregular outline. Zoom in, and this irregularity persists at smaller scales—individual mountains, ridges, and valleys maintain the same level of complexity. This self-similarity across scales, which can be mathematically modeled using fractal geometry, has allowed geologists and cartographers to generate realistic digital terrains for studies and simulations.

Both mountains and coastlines are also examples of how fractal geometry can emerge from random processes. The shape of a mountain range is sculpted by millions of years of random geological processes like erosion and uplift. Similarly, a coastline is constantly reshaped by the random actions of waves, tides, and geological activity. These examples underscore the fact that even in their randomness, natural processes often follow underlying mathematical patterns, resulting in fractal structures that are both complex and beautiful.

Crystals and Snowflakes

Crystals and snowflakes are fascinating natural occurrences of fractal patterns, exhibiting a high degree of symmetry and self-similarity. They beautifully illustrate the principles of fractal geometry and serve as brilliant examples of the intricacy and elegance of the natural world.

Crystals, including minerals and gemstones, are solid substances in which the atoms are arranged in a highly ordered repeating pattern extending in all three spatial dimensions. The fractal nature of crystals becomes apparent when we consider their formation process. Crystals grow by repeating a simple process: the addition of particles in specific, structured locations. This process happens at different scales, and the resulting form is a complex, self-similar structure that extends in three dimensions. This is most clearly seen in structures like pyrite crystals, where the cube shape is repeated at different scales.

Snowflakes offer another exquisite example of natural

fractals. Snowflakes form in clouds where water vapor condenses and crystallizes into a hexagonal lattice. As the snowflake falls to the ground, additional water vapor freezes onto the initial structure, creating intricate patterns that are symmetric and self-similar. The conditions of each snowflake's formation journey are unique, leading to the adage that "no two snowflakes are alike." Despite this variety, the self-similar pattern formation process ensures that each snowflake retains a fractal, hexagonal structure.

The study of crystals and snowflakes has led not only to a deeper appreciation of the aesthetics of nature but also to significant scientific advancements. The field of crystallography has enabled breakthroughs in materials science and solid-state physics, while the study of snowflakes has contributed to our understanding of weather patterns and water cycle dynamics. Once again, we see that the seemingly complex patterns and structures in nature can often be modeled and understood through the lens of fractals.

Fractal Patterns in the Cosmos

Fractal patterns are not confined to our terrestrial environment; they extend far into the cosmos, displaying the grandeur and intricacy of the universe. Cosmic structures, from the spiral arms of galaxies to the large-scale distribution of galaxies and galaxy clusters in the universe, exhibit remarkable fractal characteristics.

Let's begin with galaxies, the massive systems of stars, interstellar gas, dust, and dark matter that populate the universe. Spiral galaxies, like our own Milky Way, exhibit clear fractal patterns. The arms of these galaxies are not simple curves but intricate, branching structures with patterns repeating at different scales. The formation of these spiraling arms can be modeled using fractal mathematics, leading to a deeper understanding of galactic dynamics and structure.

Stepping back, we find more fractal patterns when we look at the large-scale distribution of galaxies in the universe.

Observations from powerful telescopes reveal that galaxies cluster together in intricate webs of filaments and voids. This cosmic web has a distinct fractal nature, with a self-similar clustering pattern repeating on scales of tens to hundreds of millions of light-years. Astrophysicists have turned to fractal models to describe this large-scale structure and to probe the mysteries of dark matter and the evolution of the universe.

These cosmic fractals show us that the principles of fractal geometry apply at all scales, from the minute details of a snowflake to the grand structures of the cosmos. They serve as a humbling reminder of our place in the universe and a testament to the power of fractal geometry to describe the patterns of nature. It's a realization that even as we gaze up at the night sky, we're looking at the same fundamental patterns that we can find in the world around us and even within us.

Fractals in Human Physiology

Our bodies, intricate and complex, are the greatest testimony to the prevalence of fractals in nature. Fractal patterns can be found in our lungs, our circulatory system, our nervous system, and even at the cellular level.

The human circulatory system is a perfect example of a fractal network, with a simple, repeatable pattern branching out into smaller and smaller vessels. Starting from the heart, the major arteries divide into smaller arteries, which further divide into arterioles, and finally into tiny capillaries. This fractal branching is not merely a result of random growth but a highly optimized system that ensures efficient oxygen and nutrient distribution to every cell in our body.

In our respiratory system, the fractal pattern is also evident. The trachea divides into two bronchi, each bronchus divides into smaller bronchioles, and eventually into tiny air sacs called alveoli. This fractal branching pattern ensures maximum surface area for gas exchange, packed into a compact volume.

In the realm of neuroscience, researchers have discovered

that neurons, the cells responsible for transmitting information throughout our body, form fractal patterns. The branching structure of dendrites and axons, extensions of the neurons, are fractal, optimizing communication between cells.

Even at the cellular level, we find fractals. The convoluted structure of our DNA, coiled and supercoiled, is a fractal pattern. This allows the vast length of DNA (about two meters if stretched out) to fit into a tiny cell nucleus.

Our bodies are the ultimate demonstration of how fractal geometry is not just a mathematical curiosity but an inherent part of nature and life. Fractals are a part of us, and we are a part of them.

Fractals in Weather Patterns

Meteorologists and climate scientists have long been fascinated by the chaotic, yet seemingly patterned nature of weather and climate systems. The onset of chaos theory and the discovery of fractal geometry have provided these scientists with powerful tools for understanding, predicting, and visualizing atmospheric phenomena.

Look at a satellite image of a hurricane, for instance. The swirling clouds form a spiraling pattern that is reminiscent of the Julia set, a well-known fractal. Closer examination of the cloud formations reveals smaller eddies and swirls within the larger pattern, a classic characteristic of fractals. The spiral arms are not perfect duplicates of the whole, but they do bear a striking resemblance to the larger structure.

On a smaller scale, cumulus clouds often exhibit a fractal structure, with smaller "bumps" or "towers" forming on the surface of larger cloud formations. This is due to the turbulent air flows that create and shape these clouds, and turbulence is itself a type of fractal phenomenon.

The fractal nature of weather patterns extends to how rain falls. If you plot the amount of rainfall over time or over a geographical area, you will find that the graph has a fractal structure, with periods of heavy rainfall and dry spells appearing at random, but with a pattern that repeats on different scales. This observation can be crucial for the creation of more accurate models for weather prediction.

Finally, the global climate system itself is a fractal system, with long-term climate trends (such as ice ages and interglacial periods) punctuated by shorter-term variations (like annual and decadal weather patterns). Understanding the fractal nature of these phenomena could be key to predicting and understanding how our climate is changing.

From hurricanes to cloud formations to rainfall, weather patterns provide some of the most compelling real-world examples of fractals. They illustrate how the seemingly chaotic and unpredictable can, with the right mathematical tools, be revealed to have an underlying order and pattern.

Fractals in Art and Architecture

Fractal patterns have long been a source of inspiration and fascination for artists and architects alike. The captivating nature of these intricate designs resonates with our innate appreciation of patterns and beauty. In this section, we explore how fractal patterns have influenced art and architecture throughout history and continue to do so today.

Historical Influences: Long before the formal discovery of fractal geometry, artists and architects were intuitively incorporating fractal patterns into their works. The intricate designs found in Islamic art and architecture, for instance, exhibit fractal-like properties. These tessellations and interwoven patterns are evident in the geometric decorations of mosques, palaces, and other structures, as well as in the design of traditional Persian carpets.

Similarly, Gothic architecture employed fractal-like patterns in the elaborate tracery and stonework of its cathedrals. The self-similar, branching forms of the flying buttresses and the detailed stained-glass windows reflect a deep connection to fractal geometry.

Contemporary Art and Architecture: With the advent of

modern computing and the formal study of fractals, contemporary artists and architects have been able to explore fractal patterns more deliberately and systematically. Digital artists have harnessed the power of algorithms to generate intricate, visually stunning fractal images that captivate viewers.

Artists like Jackson Pollock, who was known for his "drip paintings," have been found to incorporate fractal patterns into their work, either consciously or unconsciously. Researchers analyzing Pollock's paintings discovered that the seemingly random drips and splatters exhibit fractal-like properties, which may partially explain their aesthetic appeal.

In architecture, some contemporary designers have embraced fractals as a way to create visually engaging, organic forms that reflect the patterns found in nature. The works of architects like Frank Gehry, Zaha Hadid, and Greg Lynn incorporate fluid, curving forms that often exhibit fractal-like qualities.

Fractal patterns have also been used in urban planning, where they can help create more efficient and harmonious layouts for cities and neighborhoods. The distribution of green spaces, for example, can be optimized by considering fractal patterns, resulting in a more balanced and aesthetically pleasing urban environment.

From historical masterpieces to cutting-edge digital creations, fractals have played a significant role in shaping our artistic and architectural landscapes. The complex, self-similar patterns found in nature have captured the human imagination for centuries, and their influence on art and architecture is a testament to the enduring appeal of fractal geometry.

Examples where fractals can be found in art and architecture:

- **Islamic Geometric Design**: The intricate patterns in Islamic art and architecture often exhibit fractal characteristics. These designs, which are based on tessellations of geometric shapes, display the self-similarity

and scale invariance that are key features of fractals.
- **Gothic Architecture:** The detailed stonework, flying buttresses, and stained-glass windows of Gothic architecture often employ self-similar, branching patterns that are reminiscent of fractals.
- **Jackson Pollock's Paintings:** The drip paintings of Jackson Pollock have been found to exhibit fractal-like properties, suggesting that the artist may have intuitively incorporated fractals into his work.
- **Modern Digital Art**: With the advent of modern computers, artists have been able to use algorithms to generate complex fractal images. These digital creations often display a level of detail and intricacy that would be difficult to achieve by hand.
- **Contemporary Architecture**: Architects like Frank Gehry and Zaha Hadid have created structures with curving, organic forms that often exhibit fractal-like qualities. These designs, which are inspired by natural forms, suggest a new way of incorporating fractals into architecture.
- **Fractals in Urban Planning**: Fractal patterns can be used to create more efficient and aesthetically pleasing layouts for cities and neighborhoods.
- **Benoît Mandelbrot**: While not a traditional artist, Mandelbrot's work in developing the field of fractal geometry has inspired countless pieces of digital art. The fractals named after him, the Mandelbrot and Julia sets, are perhaps the most famous fractals and have been featured in many artistic creations.
- **M. C. Escher**: The Dutch artist M. C. Escher is famous for his intricate and mathematically inspired prints. Many of his works, such as "Circle Limit III," exhibit self-similar patterns, a key characteristic of fractals.
- **Andy Goldsworthy**: This British sculptor often works with natural materials, creating outdoor installations that reflect the patterns and structures found in nature. His works often display self-similar, fractal-like patterns.

- **Carlos Ginzburg**: Known for his "Fractal Art," Ginzburg has been creating fractal-inspired works since the 1980s. His work often explores the intersection of chaos, complexity, and fractals.
- **Roman Verostko**: Verostko is a pioneer of algorithmic art, creating software that generates artwork based on algorithms, including fractal patterns. His work represents an early example of how fractal geometry can be used in the creation of visual art.

These artists and their works show the breadth of fractal use in art and how these mathematical patterns can be employed for their aesthetic beauty as well as their conceptual depth.

Benoît Mandelbrot: The Father of Fractal Geometry

Benoît Mandelbrot, a Polish-born, French and American mathematician, is a figure whose influence transcends the traditional academic sphere. Best known for his development of fractal geometry, Mandelbrot's work has had a profound influence on a multitude of fields, including physics, finance, geology, and art.

While Mandelbrot's mathematical achievements are groundbreaking, his impact on the world of art is equally impressive. He didn't create artwork in the traditional sense, but his findings unlocked a whole new world of visual exploration. His work gave birth to an entirely new form of artistic expression: fractal art.

Fractal art is a form of algorithmic art created by calculating fractal objects and representing the calculation results as still images, animations, and media. It's a genre characterized by mathematical beauty and complexity—the art of the infinite, if you will. The most iconic of these are the Mandelbrot and Julia sets.

The Mandelbrot set is a set of complex numbers defined by a simple iterative algorithm, which, when visualized, creates a stunning fractal shape. It has been depicted in countless ways, with artists using different color schemes, zoom levels, and perspectives to highlight its intricate patterns and structures. The Julia set, meanwhile, is closely related to the Mandelbrot set and is equally captivating.

Mandelbrot's work has served as the foundation for countless digital artists, who have used his fractal geometry to create mesmerizing works of art. These artists use software to explore the Mandelbrot set (and other fractals), zooming in to find beautiful and unexpected shapes hidden within the mathematical structure.

In a very real sense, Mandelbrot himself could be considered a kind of artist. His canvas was the complex plane, and his paint was the mathematics of fractals. Through his work, he revealed a universe of beauty that was previously unknown, a universe that artists continue to explore to this day.

M. C. Escher: The Illusionist of Fractal Art

Maurits Cornelis Escher, more commonly known as M. C. Escher, was a Dutch graphic artist known for his mathematically inspired works. Born in 1898, Escher's artistic career spanned much of the twentieth century, during which he created some of the most recognizable images in modern art. While he may not have identified himself primarily as a fractal artist, his work undeniably displays a fascination with the concepts of infinity, recursion, and self-similarity that are at the heart of fractal geometry.

Escher's work often played with perspective and the representation of infinite spaces within the confines of a finite, two-dimensional plane. His creations were a masterful blend of art and mathematics, blurring the line between the physical world and the world of ideas. He had a knack for visualizing complex mathematical concepts in ways that were not just comprehensible but also aesthetically pleasing.

One of the most striking examples of this is his famous lithograph print, "Circle Limit IV (Heaven and Hell)" (1960), where he used hyperbolic geometry to depict an infinitely repeating pattern of angels and devils. The creatures decrease in size as they approach the edge of the circle, giving the impression of an infinite expanse within a finite space. This piece is a beautiful example of how Escher used concepts of self-similarity and infinity, two key characteristics of fractals, in his work.

Another example is "Drawing Hands" (1948), a lithograph that depicts two hands drawing each other into existence. This image beautifully illustrates the concept of recursion, a process where objects are defined in terms of themselves, a fundamental element of fractal geometry.

While Escher may not have had the formal mathematical training to fully grasp the principles of fractal geometry (which was still being developed during his lifetime), he nevertheless intuitively applied many of its principles in his work. His unique combination of artistic vision and mathematical intuition has left a legacy in the world of art and continues to inspire artists and mathematicians alike.

Andy Goldsworthy: The Natural Fractal Artist

Andy Goldsworthy, born in 1956, is a British sculptor, photographer, and environmentalist producing site-specific sculptures and land art situated in natural and urban settings. His work is transient, engaging directly with the environment, using natural materials found on site to create intricate, often fractal-like patterns and structures that eventually return to their natural state.

The work of Goldsworthy is characterized by his deep respect for nature and its inherent patterns and rhythms. His creations, often temporary and vulnerable to the elements, reflect the transient beauty of the natural world. However, it's the way he engages with the idea of fractals that sets him apart.

One of his most celebrated works, "River Stones," is a perfect example. In this piece, Goldsworthy cracks open a large stone and fills the inside with a collection of small stones, creating a contrast between the large, singular object and the multiple smaller objects within. This evokes the fractal property of self-similarity, where part of the object mirrors the whole.

In other pieces, Goldsworthy has created spirals from leaves, laid out in a large circle, each leaf connected to the next, the line of leaves swirling toward the center. Here, again, we see the fractal property of recursion, with the spiral pattern repeating on smaller and smaller scales.

His "Cairns" series demonstrates the concept of infinity in fractals. Goldsworthy builds stone cairns in various places around the world, each one unique, but all linked by their design and the artist's process. As each cairn is built and then erodes over time, it reflects the endless cycle of creation and decay in nature.

Goldsworthy's work is a testament to the fractal patterns inherent in the natural world. By focusing on natural and biodegradable materials, his art speaks to the impermanence of life and the beautiful complexity that can emerge from simple rules repeated over and over again. His art not only embodies the principles of fractal geometry but also reminds us of our connection to the natural world and its inherent patterns of growth, decay, and regeneration.

Carlos Ginzburg: Fractals and the Human Eye

Carlos Ginzburg, born in 1945 in Buenos Aires, is an Argentinian artist recognized for his distinctive use of fractals, particularly in his exploration of the human eye's representation. His journey into the world of fractals began in the late 1970s, when he started to experiment with repeated patterns and recursive structures.

Ginzburg's exploration of fractals was not merely a stylistic choice but a philosophical and intellectual exploration. He saw in fractals a language to express the complexity of human experience and perception, and the ability of our minds to perceive order amid chaos.

The series "Eye and Landscapes" remains one of his most celebrated works, where he employs fractal geometry to depict the landscape within the contour of an eye. Here, the fractal concept of self-similarity is interpreted as the reflection of the world in the eye—a reflection that, in turn, contains the whole world within it. This echoes the fractal notion that each part, no matter how small, contains a reflection of the whole.

In his "Fractal Cities" series, Ginzburg applied fractal geometry to create urban landscapes. The cityscapes depicted have a distinctly fractal quality, with self-similar patterns recurring at different scales, reflecting the organized chaos that characterizes urban life.

Ginzburg's work is not just an application of fractal geometry for aesthetic purposes. It's an exploration of the idea that our perception of reality is inherently fractal. Just as fractals contain infinitely complex detail, so too does our perception of the world contain layers of complexity and depth that can be explored indefinitely.

By integrating fractals into his artwork, Ginzburg provides a potent visual metaphor for the complex interplay between human perception and the world around us. His art not only uses the language of fractals to explore this complexity but also invites us to consider how fractal patterns shape our understanding of the world and our place within it.

Philosophical Reflections on Fractals in Our Surroundings
Fractals are more than just mathematical constructs or visually captivating forms; they offer us a unique lens through which to contemplate the world around us. As we've explored in the previous sections, fractals are everywhere—from the branching patterns of trees and rivers to the complex forms of mountains, clouds, and galaxies, and even in the realm of art and architecture. This ubiquity invites deep philosophical reflection.

One of the most compelling aspects of fractals is their inherent self-similarity. This concept, where the whole is reflected in its parts and each part holds within it a reflection of the whole, echoes ancient philosophical ideas. In the Western tradition, this is reminiscent of the Hermetic principle "As above, so below," which posits that patterns repeat across different levels of reality. In Eastern philosophy, it aligns with the concept of Indra's net, a metaphor used in the Avataṃsaka Sūtra to illustrate the interpenetration and interconnectedness of all things.

Additionally, fractals challenge our traditional perceptions of dimensionality. They exist in the spaces in between our established dimensions, unsettling our usual understanding of space and time. In doing so, they invite us to expand our understanding of reality itself. It can be argued that they reflect the multi-layered complexity and intricacy inherent in the natural world, suggesting that our universe might be more complex and interconnected than we can comprehend through linear or binary thinking.

The infinite complexity of fractals and their boundless intricacy unfolding from simple rules also invites reflection on the nature of infinity. This is a concept that has puzzled philosophers and mathematicians alike for centuries. Fractals bring us face-to-face with a tangible, visual form of infinity, opening new ways to engage with this elusive concept.

Furthermore, the prevalence of fractal patterns in nature has led some to propose that they represent an essential principle of growth and development. This connects to philosophical discussions about the fundamental principles that govern our universe. Are the patterns that we see in nature simply a result of chance, or do they reflect some underlying, organizing principle? Fractals, with their intricate beauty emerging from simple processes, suggest the latter.

In the realm of art, the use of fractals prompts reflection on the relationship between nature, mathematics, and human creativity. Artists like Escher and Verostko, who incorporate fractal principles into their work, demonstrate how our human desire for pattern and order aligns with the patterns and order found in nature.

In sum, the presence of fractals in our surroundings—in the natural world, in art and architecture, and even in the digital realm—invites us to see the world in a new light. They encourage us to question, to reflect, and to remain open to the complexity and intricacy of our universe. They remind us that even amid chaos and randomness, there can be order and beauty, if only we know where to look.

Let's Look at the Hermetic Principle in More Depth
The Hermetic principle, "As above, so below," is a cornerstone of Western esoteric traditions, originating from the ancient texts known as the *Corpus Hermeticum*, attributed to the mythical sage Hermes Trismegistus. This principle posits that the patterns and laws that apply to the cosmos also apply to individual beings, and vice versa. In essence, it suggests a mirroring effect, where the macrocosm (the universe) and the microcosm (the individual) reflect each other. It's a principle that points toward a fundamental unity and interconnectedness in the universe.

When we consider fractals, this Hermetic principle comes alive in a vivid and tangible way. A fractal is a geometric figure where each smaller part reflects the whole. In other words, the pattern at the micro level (the small scales) is a replication of the pattern at the macro level (the large scales). This notion of self-similarity, where the small reflects the large and the large reflects the small, is a direct visual representation of the "As above, so below" principle.

For instance, consider the branching pattern of a tree. Each branch is a smaller version of the tree itself, and each twig is a smaller version of the branch. If we zoom in on any part of the tree, we see a smaller version of the whole. This pattern continues at even smaller scales, all the way down to the veins on individual leaves. This reflects the Hermetic principle where each individual part (below) reflects the pattern of the whole (above).

From a philosophical perspective, the presence of fractal patterns in nature could be seen as evidence for the Hermetic belief in a fundamental unity and interconnectedness in the universe. It suggests that the same principles and patterns apply at all scales of existence, from the tiniest microorganisms to the vast structures of the cosmos.

Furthermore, this relationship between fractals and the Hermetic principle invites us to ponder on the nature of reality itself. If the patterns that govern the cosmos are the same as those that govern individual beings, what does that imply about our place in the universe? It suggests a deep interconnection between all things and prompts us to see ourselves not as separate entities but as integral parts of a unified whole.

In this light, fractals offer us not only a mathematical and visual marvel but also a profound philosophical symbol. They serve as a reminder of our interconnection with the universe and the underlying unity that permeates all levels of existence.

Let's Look at Fractals and Indra's Net in More Depth

Indra's net, a concept originating from the Avataṃsaka Sūtra of Mahayana Buddhism, is a profound metaphor that illustrates the interconnectedness of all things in the universe. The metaphor envisions a vast net, with each node representing an individual being or entity. At each node, there is a shining jewel that reflects all the other jewels in the net. This means that each jewel contains the reflection of all other jewels and the reflection of each jewel within those reflections, creating an infinite and interrelated web of connections.

The concept of Indra's net aligns closely with the idea of fractals in several ways. Both Indra's net and fractals express the idea that everything in the universe is interconnected. In Indra's net, each jewel contains the reflection of every other jewel, illustrating the interdependence of all beings. Similarly, in a fractal, each part of the pattern is connected to every other part, with each smaller scale mirroring the larger one.

Self-similarity: The idea of self-similarity is central to both Indra's net and fractals. In Indra's net, each jewel reflects every other jewel, creating an infinite recursion of reflections. In a fractal, the same pattern is repeated at different scales, with the smaller patterns mirroring the larger ones. This self-similarity highlights the notion that the same principles and patterns govern all levels of existence.

Infinite complexity: Both Indra's net and fractals exhibit infinite complexity. In the case of Indra's net, the infinite recursion of reflections within reflections creates an endlessly intricate web of connections. Fractals, on the other hand, can be infinitely complex, with their patterns continuing to reveal new details as we zoom in further.

From a philosophical standpoint, the presence of fractal patterns in nature resonates with the concept of Indra's net, suggesting a deep interconnectedness between all things in the universe. This interconnectedness invites us to reconsider our understanding of the world, encouraging us to see ourselves not as isolated entities but as integral parts of a vast, interconnected network.

In this sense, fractals and Indra's net both serve as powerful reminders of our interconnectedness with the universe and the underlying unity that permeates all levels of existence. By contemplating these concepts, we can gain a deeper appreciation for the intricate web of connections that bind us together and the profound beauty inherent in the patterns of the cosmos.

What Do Fractals Tell Us about Our World?

Fractals have emerged as a powerful lens through which we can understand and appreciate the complexities of the world around us. From the intricacies of our natural world to the patterns found in human-made structures and creations, fractals pervade our surroundings in ways that we might not even realize. They provide a common language that bridges the divide between the realms of mathematics, physics, biology, art, and philosophy.

The Universe's Language: Fractals are the language of the universe. They reflect a certain kind of order born out of chaos, a self-similarity across scales that we see manifested in the galaxies spiraling in the cosmos down to the microscopic patterns of our DNA. They show us that the universe is not as random as it may seem; it operates on principles that allow complexity to emerge from simple rules. This is a profound insight that has broad implications for how we understand and interpret the world.

A More Holistic Understanding: Understanding the world through the lens of fractals allows us to appreciate the complexity and diversity of nature and human creations. Fractals give us a way to quantify the seemingly unquantifiable, providing a mathematical framework to describe the irregular and complex shapes and patterns we see around us.

The concept of fractal dimensions helps us understand that dimensions are not just whole numbers but can be fractional, allowing for a more nuanced understanding of the world that incorporates various levels of complexity. This could change our perception of space and time, challenging our conventional understanding and opening new avenues of thought and exploration.

A Universal Principle: Fractals remind us of a universal principle that underlies our world: simplicity breeds complexity. The recursive process that generates fractals is based on simple rules, yet it can create incredibly complex and beautiful patterns. This principle is seen in nature and in human creations, revealing a fundamental aspect of how our world works.

Interconnectedness of All Things: Fractals reflect the interconnectedness of all things. The idea that the same patterns repeat across different scales—the essence of self-similarity—shows that everything in the universe is interconnected and interdependent, echoing philosophical thoughts from both Eastern and Western traditions. The realization of this interconnectedness encourages us to appreciate the world in its entirety, recognizing that every part, no matter how small, contributes to the makeup of the whole.

Fractals reveal a deeper, more intricate picture of our world. They illuminate the inherent complexity of the natural and cultural world, suggest a universal language of patterns, and provide a profound philosophical perspective on the interconnectedness of all things. They offer us a way to grasp infinity, reflect on the nature of reality, and appreciate the beauty and complexity that pervade our universe.

5: DEEP THOUGHTS: THE PHILOSOPHY AND ETHICS OF FRACTALS

> What if some day or night a demon were to steal after you into your loneliest loneliness and say to you: "This life as you now live it and have lived it, you will have to live once more and innumerable times more"…Would you not throw yourself down and gnash your teeth and curse the demon who spoke thus? Or have you once experienced a tremendous moment when you would have answered him: "You are a god and never have I heard anything more divine."
> —Friedrich Nietzsche, *The Gay Science*

In our exploration of fractals and their implications, we've retraced the intellectual paths of the Ancient Greeks. Pioneers like Anaximander, Socrates, and Plato pondered over the concept of infinity, a concept that surfaces repeatedly when we delve into the world of fractals.

Anaximander, with his idea of the "boundless" as the origin of all things, hinted at the unending intricacy of fractals. Socrates, as recorded in the dialogues of his student Plato, preferred a more finite universe, yet the paradoxes he used in his arguments echo the paradoxical nature of fractals—finite forms with infinite detail. Plato himself, with his world of perfect Forms, might have found the exact, self-similar nature of mathematical fractals pleasing, as an echo of the eternal Forms in the imperfect physical world.

These early thinkers paved the way for Zeno and his paradoxes, which challenged the nature of space, time, and motion—challenges that resonate with the counterintuitive nature of fractals. Zeno's paradoxes, like the concept of a line being divided ad infinitum, resonate with our understanding of fractals, whose complexity continues indefinitely, no matter how much we zoom in.

The Islamic Golden Age and The Infinite

The Islamic Golden Age, spanning from the eighth to the fourteenth century, was a time of immense intellectual growth and advancements in various fields of knowledge, from mathematics and astronomy to philosophy and medicine. The scholars of this period, standing on the shoulders of the Greeks, furthered the exploration of infinity, a concept that is central to our understanding of fractals.

One of the most notable mathematicians from this period was Persian scholar al-Khwarizmi, often regarded as the "father of algebra." He developed numerous mathematical techniques that are still in use today, some of which play a crucial role in the study of fractals. Al-Khwarizmi's algebraic methods allow us to manipulate and understand abstract mathematical structures, including those that stretch into infinity, like fractals.

Another influential figure was the philosopher and polymath al-Farabi. Known as "the Second Teacher" (after Aristotle), he made significant contributions to philosophy, logic, and metaphysics. Al-Farabi's cosmological ideas, particularly his concept of the emanation of all existence from a single, infinite source, bear echoes of the recursive, self-generating nature of fractals.

During this time, Islamic art also reflected a fascination

with the infinite. The intricate geometric patterns found in Islamic architecture, such as the endlessly repeating designs in the Alhambra Palace in Spain or the complex interlacing designs in the mosques of Iran, can be seen as physical expressions of the concept of infinity. These patterns, like fractals, repeat at various scales, creating a sense of the infinite within the finite.

In the realm of theology, the Islamic concept of God as the Infinite and the Boundless aligns with the boundless complexity and scale found within fractals. The Islamic emphasis on Tawhid, the Oneness of God, could be seen as a philosophical parallel to the self-similarity found in fractals—the idea that the whole is reflected in each part.

The Islamic Golden Age brought a new depth to the exploration of the infinite, with its advancements in mathematics, philosophy, and art. This exploration of the infinite, a concept so intrinsic to the fractal geometry we study today, has deep roots in the intellectual traditions of this period. As we navigate the intricacies of fractals, we continue a journey that was begun by these early scholars—a journey into the infinite.

Jewish Scholars of the Middle Ages and Their Views

The Jewish scholars of the Middle Ages, like their Islamic contemporaries, made profound contributions to the understanding of the infinite. They grappled with the concept in the realms of philosophy, theology, and even mysticism, with the Kabbalistic tradition.

One such scholar was Rabbi Moses ben Maimon, better known as Maimonides or the Rambam. Maimonides was a towering figure in Jewish thought, a physician, philosopher, and legal scholar who synthesized Aristotelian philosophy with Jewish theology. Maimonides's concept of God was of an absolute and infinite being, beyond human comprehension. His *Guide for the Perplexed* is a deep dive into the nature of the divine and the infinite. It's not exactly light reading, though! One story goes that a student once complained about the difficulty of his work. Maimonides is said to have responded, "You think it's difficult to read? You should try writing it!"

Maimonides also used the concept of potential and actual infinity, a precursor to the concept used in calculus and in turn, in fractal mathematics. In his view, the universe was potentially infinite but actually finite, a concept we find mirrored in the finite but unbounded nature of fractals.

In the realm of mysticism, the Kabbalistic tradition provides fascinating perspectives on the infinite. The concept of *Ein Sof*, or "without end," represents the divine as an infinite entity. The ten *sefirot*, or divine emanations, are portrayed as interconnected nodes in the Tree of Life, a diagram that bears a striking resemblance to fractal structures.

The Kabbalists even had a sense of humor about the infinite. One anecdote tells of a Kabbalist who claimed he could see from one end of the world to the other using a special lens. When asked to prove it, he said, "Sure, just as soon as you can show me where the world ends!"

The Jewish scholars of the Middle Ages thus brought their unique perspectives to the exploration of the infinite, a concept that is integral to our modern understanding of fractals. Their philosophies, theological perspectives, and yes, even their humor, continue to illuminate our understanding of the infinite today.

Maimonides and His Guide for the Perplexed: A
Connection to Infinity and Fractals

Rabbi Moses ben Maimon, known as Maimonides or the Rambam, was a significant figure in the Middle Ages who tackled the vast concept of infinity in his seminal work, *The Guide for the Perplexed*. This book is a philosophical treatise intended to reconcile Aristotelian philosophy with Jewish religious thought, and it has had a profound influence on both Jewish and non-Jewish thought for centuries.

One of the central themes in Maimonides's work is the nature of God and the infinite. According to Maimonides, God is the absolute and ultimate form of infinity—beyond all physical limitations, beyond time, and even beyond comprehension. He argued that human language and thought are too limited to fully comprehend the infinite nature of God. This theme can resonate with our modern understanding of fractals, which are themselves infinite, complex, and often beyond intuitive understanding.

There's a story that Maimonides, while writing *The Guide for the Perplexed*, was approached by a student who was struggling to comprehend the text's complexities. The student said, "Rabbi, your book is titled 'Guide for the Perplexed,' but after reading it, I am even more perplexed!" Maimonides smiled and responded, "It is not a 'Guide for Those Who Are Not Yet Perplexed.' Only by being perplexed, by questioning and grappling with difficult concepts, can we begin to understand."

Maimonides also presents the concept of potential and actual infinity in his philosophical system, a concept closely related to fractal mathematics. He proposed that the universe could be potentially infinite but is actually finite. This mirrors the concept of fractals, which can be infinitely complex but contained within a finite space.

Thus, Maimonides, with his deep philosophical insights and engaging anecdotes, provides a unique lens to explore the concept of infinity, offering parallels to our understanding of the infinite complexity and potential of fractals.

St. Augustine and the Christian Fathers: Fractals and the Infinite
St. Augustine, one of the most influential Christian theologians, made significant contributions to the understanding of infinity, and his insights could also be linked to the conceptual universe of fractals.

Augustine lived in the fourth and fifth centuries AD, a time of great upheaval and change in the Roman Empire. Amid this turmoil, he pondered deeply about the nature of God, time, and the eternal. His thoughts on these matters were recorded in his seminal work, *Confessions*, where he grappled with the nature of God and eternity.

The concept of God in Augustine's philosophy is inherently infinite. God is depicted as a being without boundaries or limits, a being of infinite power, knowledge, and presence. This depiction resonates with the concept of fractals, which, despite their finite physical manifestation, contain infinite detail and complexity within their structures.

Moreover, Augustine's exploration of time can also be interpreted through the lens of fractals. In *Confessions*, he famously asked: "What then is time? If no one asks me, I know what it is. If I wish to explain it to him who asks, I do not know." This paradoxical nature of time, known yet elusive, can be seen as a fractal characteristic. Just as a fractal pattern repeats and varies across different scales, time also seems to have a self-similar nature, with the past, present, and future continually echoing each other.

Augustine's philosophical ideas, particularly his understanding of God and time, provide a unique perspective on the fractal-like nature of reality. His struggle to understand and articulate the infinite and the eternal resonate with our modern mathematical efforts to comprehend the infinite complexity of fractals.

While the Christian fathers did not have the mathematical language of fractals, their philosophical and theological insights can offer us a deeper understanding of the infinite complexity inherent in fractals and the paradoxes that come with grappling with the infinite.

Medieval Philosophers and the Infinite

The Middle Ages, a period stretching from the fifth to the fifteenth century, was a time of philosophical flourishing in Europe. It was during this period that several philosophers made profound contributions to the understanding of the infinite, which, in retrospect, can be related to the concept of fractals.

Thomas Aquinas and the Infinite: Thomas Aquinas, a prominent figure of medieval scholasticism, grappled extensively with the concept of the infinite in his philosophical and theological works. Aquinas argued that while God is indeed infinite, actual infinite quantities cannot exist in the physical world. He asserted that the actual infinite is a concept reserved for God alone.

However, Aquinas accepted the existence of potential infinite. For instance, time, he said, is potentially infinite; it does not have an actual infinite quantity but can continue indefinitely. This mirrors our understanding of fractals, as they possess the property of self-similarity over an infinite range of scales; in a sense, a form of potential infinity.

Duns Scotus and the Infinite: John Duns Scotus, another influential medieval philosopher, made significant contributions to the debate on the infinite. Unlike Aquinas, Duns Scotus argued for the existence of the actual infinite in the physical world. He proposed that there could be an actual infinite number of "simultaneously existing individuals," a concept that parallels the infinitely complex and detailed structure found in fractals.

William of Ockham and the Infinite: William of Ockham, famous for the "Ockham's razor" principle, was another medieval thinker who wrestled with the notion of the infinite. Ockham's approach to infinity was pragmatic. He accepted the idea of potential infinity, like Aquinas, and rejected the existence of the actual infinite in nature.

***

The medieval philosophers' exploration of the infinite provides a philosophical backdrop against which we can consider the concept of fractals. While these thinkers did not have the mathematical framework of fractals at their disposal, their ideas resonate with modern fractal theory's understanding of infinite complexity and detail.

The Italian Renaissance and the Infinite

The Italian Renaissance, a period stretching from the fourteenth to the seventeenth century, was a time of renewed interest in the cultures of ancient Rome and Greece. This era saw a surge in artistic and intellectual activity, leading to a significant shift in the understanding of the infinite, which can be linked retrospectively to the concept of fractals.

Leonardo da Vinci and the Infinite: Leonardo da Vinci, one of the most celebrated figures of the Renaissance, had a profound interest in nature and its intricate patterns. He studied and sketched natural forms, such as the branching patterns of trees and the spiral designs of shells. Although he didn't have the mathematical framework of fractals, his observations align with the notion of self-similarity at different scales—a fundamental property of fractals.

Luca Pacioli and Divine Proportion: Luca Pacioli, a contemporary of Leonardo, wrote extensively on mathematics and is best known for his work, *Divina Proportione* (*Divine Proportion*). This work explores the golden ratio, a proportion that is self-replicating and infinite, properties reminiscent of fractals. While Pacioli didn't specifically address the concept of the infinite in his works, his exploration of the golden ratio underlines the notion of infinite self-similarity, a key feature of fractals.

Giordano Bruno and the Infinite Universe: Giordano Bruno, a philosopher and cosmologist, put forth the radical notion of an infinite universe filled with a potentially infinite number of worlds. His concept of infinity extended beyond the theological domain and into the physical realm, offering a new way to comprehend the infinite in the universe. This idea resonates with the fractal perspective, which sees infinity not just as an abstract concept, but as a property inherent in the physical world.

***

In conclusion, the Italian Renaissance's examination of the

infinite provides a fascinating historical context to the contemporary understanding of fractals. The thinkers of this era, much like the philosophers of the Middle Ages, did not have access to the mathematical framework of fractals, yet their ideas reverberate with the infinite complexity and detail intrinsic to fractal theory.

The Mathematicians of the Italian Renaissance

The Renaissance in Italy was a time of extraordinary intellectual and artistic activity. The rebirth of classical learning ignited a surge of interest in mathematical studies, laying the groundwork for future advances in this field. While the concept of fractals was not yet known, the work of several mathematicians during this era would prove crucial to its future development.

Luca Pacioli: As previously mentioned, Luca Pacioli was a major figure in the Italian Renaissance. Known as the father of accounting and bookkeeping, he wrote one of the first comprehensive books on mathematics, *Summa de arithmetica, geometria, proportioni et proportionalita* in 1494. His work *Divina Proportione*, which explores the golden ratio, is particularly relevant to the study of fractals.

Gerolamo Cardano: Gerolamo Cardano was an Italian polymath who made significant contributions to algebra. In his 1545 book *Ars Magna*, he published solutions to the cubic and quartic equations, which included the concept of complex numbers. Complex numbers would later play a key role in the development of fractal geometry, particularly in the creation of the Mandelbrot set.

Niccolò Fontana Tartaglia: Niccolò Fontana Tartaglia was another notable mathematician of this era. He is best known for his solution to cubic equations, which he published in his 1535 book *Nova Scientia*. His work was significant for its exploration of patterns and symmetries in numbers, a theme which is central to the study of fractals.

Rafael Bombelli: Rafael Bombelli was an Italian mathematician known for his work on complex numbers. His 1572 book

Algebra was the first to systematically use complex numbers and provide rules for their manipulation. This foundational work on complex numbers would be critical to the later development of fractal geometry.

The mathematicians of the Italian Renaissance, while not directly contributing to the study of fractals, laid important groundwork for later developments in this field. Their exploration of mathematical concepts such as the golden ratio, complex numbers, and patterns in numbers paved the way for the eventual discovery of fractal geometry.

The Age of Enlightenment and Beyond

The Age of Enlightenment in the seventeenth and eighteenth centuries was a time of great intellectual and scientific development. Several key ideas concerning infinity and the nature of the universe arose during this period, which would later contribute to our understanding of fractals.

Blaise Pascal: Blaise Pascal, a French mathematician, physicist, and philosopher of this period, offered profound insights into the nature of infinity. In his collection of thoughts, known as *Pensées*, he wrote about the vastness of the universe and the infinitesimal place of the human within it. Pascal's paradox, the notion that "the infinite and the finite are equally incomprehensible," aligns with the concept of fractals, as they are finite shapes that contain infinite complexity.

Gottfried Wilhelm Leibniz: German philosopher and mathematician Gottfried Wilhelm Leibniz had a unique perspective on infinity. He was one of the first to use the term "fractal" in a mathematical context, although his use of the term differed from its modern meaning. Leibniz also developed the concept of monads, which he described as simple substances that make up the universe. These monads, according to Leibniz, reflect the entire universe in a way that is reminiscent of the self-similarity found in fractals.

Georg Cantor: Moving into the nineteenth century, Georg Cantor made significant contributions to the understanding of infinity. He established the concept of different sizes of infinity and proved that there are as many points in a one-inch line segment as there are in a line of infinite length. This idea of differing infinities resonates with fractal geometry, where an infinitely complex structure can exist within a finite space.

Benoît Mandelbrot: Finally, we arrive at the twentieth century with Benoît Mandelbrot, the father of modern fractal geometry. Mandelbrot built on the ideas of his predecessors, harnessing the power of modern computers to explore the complexity of fractals. He coined the term "fractal" to describe shapes that are self-similar at different scales and used this concept to explain complex natural phenomena.

***

Thus, through the ages, philosophers and mathematicians have grappled with the concept of the infinite and the complex. The development of fractal theory is deeply rooted in this historical exploration, reflecting humanity's persistent quest to understand the universe's complexity and our place within it.

Leibniz and the Birth of Fractal Thought

Gottfried Wilhelm Leibniz, a remarkable figure in the history of philosophy and mathematics, significantly contributed to the early roots of fractal thought. His extraordinary intellect and insatiable curiosity gave rise to many fascinating stories and anecdotes.

Leibniz was born in 1646 in Leipzig, Germany. Known for his talent in varied disciplines, he is particularly famous for developing calculus independently of Sir Isaac Newton, leading to a heated priority dispute between the two.

In his metaphysical work, Leibniz introduced the concept of monads, which he described as simple substances that make up the universe. These monads, according to Leibniz, reflect the entire universe in a way that is reminiscent of the self-similarity found in fractals. This fascinating idea suggests that each part of the universe, no matter how small, mirrors the complexity and intricacy of the whole, an idea central to the concept of fractals.

One amusing anecdote about Leibniz concerns his invention of a mechanical calculator, known as the Stepped Reckoner, which could perform all four basic arithmetic operations. It's said that when Leibniz demonstrated the device to Peter the Great of Russia, the Tsar was so intrigued that he climbed onto a table to get a better look. Unfortunately, the table collapsed under his weight, leading to a rather comical scene with the Tsar sprawled amid the wreckage of the table and the disassembled calculator.

Leibniz was also known for his optimism, summed up in his statement that we live in the "best of all possible worlds." This outlook was famously satirized by Voltaire in his work *Candide*, in which Dr. Pangloss (a character representing Leibniz) continually insists that everything happens for the best, no matter how dire the circumstances.

These stories illustrate Leibniz's unique character: a genius inventor with a positive outlook, unafraid of challenging established ideas. His work planted early seeds for the concept of fractals, and he serves as an inspirational figure for all those fascinated by the intricate and infinite patterns found in nature and mathematics.

Leibniz's Monadology and Its Connection to Fractals: Leibniz's concept of monads in his work *Monadology* presents a fascinating parallel to the fractal nature of the universe. Monads, according to Leibniz, are the fundamental, indivisible elements of the universe. Each monad is essentially a mirror of the entire universe. In other words, every part contains a reflection of the whole. Despite being simple and "windowless," each monad represents a microcosm of the entire cosmos. This conception of monads encapsulates an idea that is strikingly similar to the concept of self-similarity in fractals, where each part of a fractal reflects the entire structure.

The concept of self-similarity is central to fractals. This means that a fractal pattern, no matter how much you zoom in or zoom out, always retains the same level of detail. Fractals, therefore, are infinite, just like the reflections contained within each of Leibniz's monads.

Furthermore, the monads' interrelatedness in Leibniz's philosophy is similar to the interconnectedness in a fractal structure. In a fractal, the change in one part affects the entire structure, due to its iterative nature. Similarly, for Leibniz, a change in one monad reflects in all monads, as they are all mirrors of the same universe.

In a sense, Leibniz's monads can be thought of as philosophical fractals. They represent a universe where every part, no matter how minute, contains a reflection of the whole. Just as in a fractal, the complexity and intricacy of the whole are present at every level. This connection between Leibniz's monads and fractals presents an intriguing blend of philosophy and mathematics, showcasing the recurring theme of infinity and self-similarity across different fields of thought.

The Nature of Monads: Monads, as conceptualized by Leibniz, are fundamental units of reality. They are elementary particles, so to speak, but not in the physical sense. Rather, they are metaphysical entities. Leibniz presents monads as simple substances, devoid of parts, and hence, indivisible. They are the ultimate elements that constitute the fabric of the universe. As basic building blocks, they exist independently, without any need for space or time. This means they are also timeless and indestructible, making them infinitely enduring entities.

Leibniz also characterizes monads as "windowless." By this, he means that monads have no interaction with the outside world. They have no doors or windows through which something might enter or exit. This may seem counterintuitive given their role as the constituents of a dynamically interconnected universe, but Leibniz has a unique way of reconciling this, which we'll delve into in the next sections.

Interestingly, these characteristics of monads parallel

certain aspects of fractals. Like monads, fractals are also indivisible in the sense that they retain their identity no matter how much you zoom in or out, reflecting an infinite simplicity within complexity. Furthermore, the idea of monads being windowless mirrors the self-contained nature of fractals, wherein each part encapsulates the whole without direct interaction with its surroundings.

With this understanding of the nature of monads, we can now explore how they act as mirrors of the universe, an aspect that further strengthens their connection to fractals.

Monads as Mirrors of the Universe: Leibniz presents a fascinating aspect of monads despite their "windowless" character: they represent or mirror the entire universe. Each monad, in its own way, contains a reflection of everything else that exists. This is a profound idea that parallels the concept of holography, where each portion of a hologram contains the entire image.

This concept also aligns with the defining characteristic of fractals—self-similarity. Just as each tiny portion of a fractal mirrors the whole, each monad mirrors the entire universe. This notion of mirroring also connects with the concept of recursion, a key principle in generating fractal patterns.

Leibniz describes this mirroring aspect of monads through the concept of perception. According to him, each monad perceives the universe in its own way, from its own perspective. However, this perception is not conscious awareness but a kind of internal representation or reflection of the external world.

Leibniz's concept of appetition further explains how monads change their perceptions over time. Appetition is an inherent tendency in monads that drives them toward the future state of their perceptions, thereby allowing them to evolve while mirroring the universe's unfolding changes.

Thus, in Leibniz's philosophy, monads serve as individual, self-contained universes, continuously reflecting the entire cosmos within themselves. They are like countless mirrors, each capturing the whole universe from a unique angle, together creating a dynamic, multidimensional mosaic of reality—a notion that resonates deeply with the infinite complexities found in fractal geometries.

Harmony in the *Monadology*—a Fractal Universe: Leibniz's *Monadology* also presents a concept of "preestablished harmony." According to this concept, all monads, though separate and without interaction, are in perfect harmony with each other. This harmony is preestablished by God, who, in Leibniz's philosophical system, is the ultimate monad, the original uncaused cause, and the source of all existence.

Leibniz suggests that each monad evolves according to its internal principles, reflecting the universe in a way that is perfectly in sync with all other monads. There is no direct causality between them—their harmony is preprogrammed, like an incredibly complex orchestral piece in which each instrument plays its part independently, yet together they produce a beautiful and coherent symphony.

This preestablished harmony echoes the fractal nature of reality, where every part, though independently existing, reflects the whole. Like a fractal, the universe, according to Leibniz, is an interconnected whole where every piece is intimately linked to the entire structure, not through direct causation, but through intricate patterns of harmony that pervade the cosmos.

This concept also suggests that the universe's complexity and order are not a result of simple mechanical interactions but of an inherent, preestablished harmony. It's an idea that challenges reductionist views of the world and invites us to embrace a more holistic, fractal understanding of reality.

Through the lens of fractal philosophy, we can see Leibniz's monads as infinite points of reflection, each containing a universe within, each evolving in preestablished harmony, mirroring the fractal nature of reality in its own unique way. In this sense, Leibniz's *Monadology* can be seen as a philosophical precursor to the idea of fractal geometry.

Ethics and Morality in Leibniz's *Monadology*—a Fractal Perspective: One of the less immediately apparent connections between Leibniz's philosophy and fractal theory is the moral and ethical implications of his *Monadology*. Remember, each monad, according to Leibniz, is a self-contained universe that mirrors the entire cosmos. This includes not only physical characteristics but also mental, moral, and spiritual ones.

From a moral perspective, the implication is profound. If each monad is a reflection of the whole, then each individual entity—whether human or not—has inherent worth and dignity. This outlook has substantial implications for ethics and morality, suggesting an innate respect for all beings and things.

In fractal terms, this is akin to understanding that every small segment of a fractal pattern carries the same weight and importance as the whole. It calls for an appreciation of the interconnectedness and interdependence of all things and beings, a key ethical insight in many spiritual and philosophical traditions.

Furthermore, if we consider Leibniz's principle of "the identity of indiscernibles," which postulates that no two distinct things can have all their properties in common, we can see a form of individual uniqueness that aligns with the infinite diversity found within fractal patterns. Like fractals, where every pattern is unique and nonrepeating, each monad carries a distinctness that's valued and respected.

So in a sense, Leibniz's philosophy offers a moral framework that's remarkably harmonious with the fractal understanding of the universe. It emphasizes respect for diversity, uniqueness, and individual worth—principles that echo loudly in our contemporary ethical debates.

The Best of All Possible Worlds—Optimism in Leibniz's *Monadology* and Fractal Geometry: In addition to offering a novel way to understand the world, Leibniz's *Monadology* also presented a rather optimistic outlook on life and the universe. Leibniz famously declared that we live in "the best of all possible worlds," a statement that has been the subject of much philosophical debate and discussion.

In Leibniz's view, God, being omniscient and omnipotent, created a world where the maximum amount of variety and complexity is packed into the simplest laws of nature. The result is a universe that, despite suffering and evil, is optimized for the greatest possible good.

Fractal geometry, interestingly, echoes this sentiment of "optimization." Fractals are patterns that can generate infinite complexity from simple rules, much like how Leibniz viewed the world. They're a testament to the incredible efficiency and elegance of nature, where simple laws can give rise to the intricate beauty and diversity we see around us.

Moreover, the recursive and self-similar nature of fractals can be seen as a metaphor for the interconnectedness and unity of all things, which is a central theme in Leibniz's philosophy. Just as every part of a fractal reflects the whole, each monad in Leibniz's system is a microcosm that mirrors the entire universe.

Thus, the optimism inherent in Leibniz's *Monadology* and the elegance of fractal geometry seem to intertwine beautifully. Both hint at a world that, despite its apparent chaos and complexity, is underpinned by simple, elegant laws and is, in some profound sense, the best it can possibly be. This viewpoint can provide a comforting and positive perspective in our modern, often chaotic, world.

Cantor's Set Theory

Georg Cantor, a German mathematician, made significant contributions to set theory, a branch of mathematical logic dealing with collections of objects, during the late nineteenth century. His work fundamentally changed how mathematicians understand the concept of infinity, laying the groundwork for much of modern mathematics.

Before Cantor, the concept of infinity was largely unexplored in rigorous mathematical terms. The Greek philosopher Zeno, for example, posed paradoxes involving infinite sequences, but these remained philosophical curiosities without a formal mathematical framework.

Cantor's revolutionary idea was to treat infinity as a valid size for a set, thus introducing the concept of an "infinite set." More strikingly, Cantor proposed that there could be different sizes of infinity. This idea was encapsulated in his concept of "cardinality," which refers to the number of elements in a set.

In Cantor's framework, two sets have the same cardinality if their elements can be paired off one to one with no elements left over in either set. Using this definition, Cantor showed that the set of natural numbers (1, 2, 3…) has the same cardinality as the set of even numbers, even though the latter is a subset of the former. Both sets, he argued, are "countably infinite."

In contrast, Cantor showed that the set of all real numbers between zero and one is "uncountably infinite," meaning that it cannot be put into a one-to-one correspondence with the natural numbers. This was a profound and counterintuitive result, suggesting that some infinities are larger than others.

Cantor's groundbreaking work on set theory and infinity was initially met with resistance in the mathematical community, but it has since become a cornerstone of the subject. The deep understanding of infinity that it facilitated is essential to many areas of mathematics, including calculus, topology, and the study of fractals.

Cantor's Discovery of Fractals: Cantor's groundbreaking work on set theory and infinity also led him to an early discovery of fractals. His Cantor set, or Cantor dust, introduced in 1883, is one of the earliest examples of a fractal.

The Cantor set is constructed by starting with the closed interval [0, 1] on the real number line and repeatedly removing the open middle third of each remaining piece. The first step removes the interval (1/3, 2/3), leaving two pieces [0, 1/3] and [2/3, 1]. The next step removes the middle thirds of these pieces, and so on. After infinitely many steps, the Cantor set is what remains.

The Cantor set is a fractal because it is self-similar: each piece looks like a smaller copy of the whole. It also has an interesting property related to its dimension. While the Cantor set is a subset of the one-dimensional real line, it is so spread out that it doesn't seem to have dimension one. Yet it is too disconnected to have dimension zero. This paradox led to the concept of fractional, or fractal, dimension.

Cantor's work on the Cantor set was part of a larger effort to understand the different possible sizes of infinity. The Cantor set, despite being derived from a finite interval by removing parts, is uncountably infinite. Moreover, it has the same cardinality as the interval [0, 1] from which it was created. This is another demonstration of the counterintuitive nature of infinity in Cantor's framework.

While the Cantor set is relatively simple to define, it has complex and surprising properties that exemplify the richness and depth of the world of fractals. It is a testament to Cantor's genius that he discovered such a mathematical object long before the word "fractal" was even coined.

Cantor's Impact and Reception: Georg Cantor faced significant opposition and criticism from some of his contemporaries. His radical ideas about infinity and his creation of set theory were not immediately accepted by the mathematical community. Some, like Leopold Kronecker, were vocal in their disapproval. Kronecker famously said, "I don't know what predominates in Cantor's theory—philosophy or theology, but I am sure that there is no mathematics there."

Despite this resistance, Cantor remained resolute in his beliefs and continued to develop his theories. His work, including his creation of the Cantor set, would later be recognized as fundamental to the development of modern mathematics. His pioneering ideas about infinity and set theory would lay the groundwork for the development of several branches of mathematics, including measure theory, topology, and indeed, fractals.

In terms of the philosophical implications of Cantor's work, his theories challenged traditional notions of infinity and introduced the concept of multiple infinities of different sizes. This was a major departure from the prevailing view of infinity as a singular, unquantifiable concept. His work thus had profound implications not only for mathematics but also for philosophy and our understanding of the universe.

In the context of fractals, Cantor's work is of paramount importance. His Cantor set was one of the earliest examples of a mathematical object with noninteger dimension, a concept that would later be integral to the definition of fractals. Cantor's work on set theory and infinity paved the way for the development of the mathematical tools necessary to study fractals, and his legacy can be seen in the rich and complex world of fractal geometry that we know today.

Cantor's Personal Struggles and Legacy: Georg Cantor's revolutionary ideas and the controversy they ignited took a heavy toll on his mental health. He suffered from bouts of depression and was hospitalized several times during his life. Despite these personal struggles, he never gave up on his work, and his theories would eventually be recognized as groundbreaking contributions to the field of mathematics.

His work, especially in the realm of set theory and the concept of different sizes of infinity, has had a profound influence on mathematics, philosophy, and theoretical physics. It laid the groundwork for the mathematical exploration of the concept of infinity and made possible the development of many modern mathematical theories and concepts, including the concept of fractals.

Cantor's legacy is thus an enduring one. His ideas, which were once considered controversial and even heretical, are now fundamental to our understanding of mathematics and the nature of infinity. His life and work are testament to the power of persistence in the face of adversity and the pursuit of truth despite societal resistance.

In the context of fractals, Cantor's contributions cannot be overstated. His pioneering work in set theory and his introduction of the concept of different sizes of infinity were instrumental in the development of fractal geometry. His bold exploration of infinity and the concept of noninteger dimensions laid the groundwork for the rich and complex world of fractals. Cantor's life and work serve as a reminder of the beauty and complexity that can arise from simple mathematical concepts and the boundless possibilities that exist within the realm of mathematical exploration.

Cantor's Influence on Fractal Geometry: Cantor's contributions to the field of mathematics, particularly his work on set theory and the concept of different sizes of infinity, had a profound impact on the development of fractal geometry.

The Cantor set, a simple yet paradoxical construct, exemplifies this influence. It is a set that is uncountably infinite yet has no length, embodying Cantor's exploration of the abstract concept of infinity. It is a perfect example of a fractal, having a noninteger fractal dimension and self-similarity across scales. The Cantor set introduced the mathematical world to the idea that objects could be broken down into an infinite number of parts and yet be devoid of substance, a concept that is at the core of many fractal constructs.

Additionally, Cantor's work on transfinite numbers and different sizes of infinity informs the way we measure and understand the complexity and dimensions of fractals. His work on countability and uncountability helps us understand the seemingly paradoxical concept that fractals can have infinite detail yet be contained within a finite space.

Indeed, Cantor's influence is deeply woven into the fabric of fractal geometry. His groundbreaking ideas about infinity, his innovative set theory, and his indomitable spirit in the face of criticism and personal struggles all contribute to his enduring legacy in the realm of fractals. His work continues to inspire mathematicians and fractal researchers, encouraging them to push the boundaries of our understanding of complexity, infinity, and the very nature of reality.

Mandelbrot's Early Life and Education

Benoît B. Mandelbrot was born in Warsaw, Poland, in 1924, into a family with a strong academic lineage. His uncle, Szolem

Mandelbrojt, was a mathematician who spent a significant part of his career in Paris, France. Due to the increasing tensions in Europe, Mandelbrot's family emigrated to France when he was just eleven years old.

In France, Mandelbrot was introduced to mathematics through his uncle, and this ignited a passion that would shape the rest of his life. Despite the disruptions caused by World War II, Mandelbrot persevered with his education. He studied at the École Polytechnique, where he was mentored by some of the most brilliant minds in the field of mathematics. Afterward, he moved to the United States to pursue further studies and obtained his PhD in mathematical sciences from the University of Paris in 1952.

During his formative years, Mandelbrot was exposed to a wide range of ideas, both within and outside mathematics. This broad intellectual background would later play a crucial role in his ability to see patterns and connections where others did not. It was these connections that eventually led him to the discovery of fractals.

Mandelbrot's Career and the Discovery of Fractals: After his education, Mandelbrot embarked on a career that straddled both academia and industry. His first significant position was at the Institute for Advanced Study in Princeton, where he rubbed shoulders with some of the greatest minds of the time. However, it was his subsequent role at IBM's Thomas J. Watson Research Center that truly set the stage for his groundbreaking work on fractals.

Mandelbrot's job at IBM, which began in 1958, involved using mathematics to solve practical problems. It was here that he started investigating peculiar patterns in seemingly random phenomena such as noise in telephone lines and the distribution of large and small galaxies in the universe. He noticed a common characteristic in these phenomena: they exhibited similar patterns at different scales, a property he called "self-similarity."

In the late 1970s, while exploring the mathematics of Julia sets (a concept developed by French mathematicians

Gaston Julia and Pierre Fatou), Mandelbrot discovered a set of complex numbers that, when iteratively applied through a simple formula, produced a shape with an infinitely intricate boundary. This shape, now known as the Mandelbrot set, became the poster child for fractals. It was a striking visual demonstration of the concept of self-similarity, as each small section of the boundary, when magnified, revealed a complexity mirroring the whole set.

His discovery was not just a mathematical curiosity. It provided a new way to describe many natural phenomena that didn't fit well into traditional Euclidean geometry. Fractals turned out to be an effective way to describe the irregular shapes and processes found in nature, such as the branching of trees, the jaggedness of coastlines, and the clustering of galaxies.

Fractals in the Digital Age: The advent of digital technology and computers provided the ideal platform for the exploration and application of Mandelbrot's fractal geometry. With their ability to perform complex calculations and render intricate graphics, computers made it possible to delve deeper into the world of fractals than ever before.

A milestone in this digital exploration was the creation of the Mandelbrot set. This set, named after Mandelbrot himself, is a collection of complex numbers that remains bounded when iterated through a specific function. The visualization of the Mandelbrot set produces stunning, infinitely complex patterns that epitomize the beauty and depth of fractal geometry.

Beyond its aesthetic appeal, the Mandelbrot set offers profound insights into the nature of complexity and infinity. Zooming into the set reveals an endless array of detail, with smaller copies of the whole shape appearing at every scale. This property, known as self-similarity, is a hallmark of fractal objects.

The impact of Mandelbrot's work in the digital age extends well beyond mathematics and science. Fractal geometry has found its way into computer graphics and digital art, contributing to the realistic rendering of natural objects like mountains, clouds, and trees in video games and movies.

Moreover, the digital visualization of fractals has played a significant role in popularizing mathematics and making it more accessible to the general public. The mesmerizing beauty of fractal images has captivated audiences worldwide, sparking interest in the underlying mathematical concepts. In this way, Mandelbrot's legacy continues to inspire and educate, demonstrating the awe-inspiring complexity and beauty inherent in the mathematical fabric of our universe.

Legacy of Mandelbrot and Fractals in Contemporary Thought: Mandelbrot's work has had a profound impact not only on mathematics, but also on our understanding of the universe and our place in it. By revealing the ubiquitous presence of fractals in nature, he has profoundly influenced fields ranging from physics to finance, from geology to music, and from computer graphics to philosophy.

In physics, for example, fractal geometry has been used to understand and model complex phenomena such as turbulence and galaxy formation. In finance, it has been applied to model the erratic behavior of financial markets, which traditional models have often failed to predict accurately. In geology, it helps us understand the complex structures of mountains and coastlines. Even in music, the concept of self-similarity has been used in the creation of certain types of music and sound synthesis.

Mandelbrot's fractal geometry has also had profound philosophical implications. It has challenged our traditional Euclidean conception of space and time, suggesting that the universe might be more accurately described as a fractal, with self-similar patterns recurring at all scales.

Mandelbrot's work on fractals has also brought about a renewed appreciation for the beauty and complexity of mathematics. The stunning visualizations of the Mandelbrot set and other fractals have made abstract mathematical concepts more accessible and engaging to the general public.

As we move further into the twenty-first century, it

is clear that the influence of Mandelbrot and his fractals will continue to resonate across diverse fields of study. The exploration of the infinite complexity of the fractal universe is just beginning, and the legacy of Mandelbrot ensures that this journey will continue to captivate and inspire future generations of thinkers and explorers.

Fractals, Ethics, and the Universe: A New Perspective
The study of fractals and their implications reaches far beyond the realm of mathematics and physics. Fractals give us a lens to explore not just the physical structure of the universe, but also the ethical and philosophical implications of living in such a universe. The works of great thinkers like Leibniz, Cantor, and Mandelbrot have laid the foundation for our understanding of fractals, and their ideas can be extended to explore the ethical dimensions of a fractal universe.

The Aesthetics of Sameness: At first glance, the nature of fractals might seem to endorse a lack of diversity—after all, they're based on the repetition of the same pattern at different scales. However, it's essential to note that this "sameness" is not uniformity. Instead, it's a testament to the immense complexity and diversity that can emerge from a single, simple rule. This idea can be a powerful philosophical tool, teaching us about the profound depths that can be discovered in exploring a single concept thoroughly. Philosophically, it leads us to question our emphasis on diversity as inherently beneficial and ask whether there might be a beauty and profundity in sameness that we overlook.

Harmony and Balance: Fractals embody a perfect balance between order (the repeated pattern) and chaos (the infinite complexity that arises from this repetition). This idea resonates with many philosophical and spiritual traditions, which emphasize the need for balance in life, whether between work and rest, giving and receiving, or other dichotomies. The fractal view of the universe suggests a cosmos that is neither wholly deterministic nor entirely random but a mixture of both.

Ethics of Interconnection: Fractal patterns are inherently interconnected, with each part related to every other. This perspective mirrors an ethical view that emphasizes the interconnectedness of all beings and the importance of considering our actions' wider effects. It aligns with the idea that we should strive for an ethical approach that acknowledges and respects our interconnectedness, as seen in philosophies like Ubuntu in Africa, which emphasizes community and mutual assistance.

Eternal Becoming: Fractals also embody the concept of eternal becoming or perpetual change. Each iteration of a fractal pattern can be seen as a transformation of the previous one, in a never-ending process of becoming. This concept echoes the philosophies of Heraclitus and other thinkers who viewed reality as a constant flux.

By providing a new lens through which to view the universe, fractals can enrich our philosophical and ethical perspectives. They invite us to explore the beauty in sameness, the richness in balance, the ethics of interconnection, and the wonder of eternal becoming.

Fractals: A New Lens to Understand Reality

Fractals give us a different way to interpret and understand the world around us. These intricate patterns, with their endlessly repeating structures at ever smaller scales, are not merely mathematical curiosities. They offer insights into the deepest truths of our existence and reality.

Fractal geometry allows us to recognize and celebrate the complex beauty of sameness. This perspective can help us appreciate the richness of experiences, ideas, and phenomena that we might otherwise dismiss as monotonous or uninteresting. It teaches us that there is an inherent richness and depth in exploring and understanding the nuances of a single entity or concept.

Fractals also suggest that our universe is built on a delicate balance between order and chaos. This idea invites us to strive for a similar balance in our lives and society, acknowledging the necessity of both predictability and unpredictability, stability and change, rigidity and flexibility.

The interconnectedness evident in fractal structures underscores the reality of our existence. We are all part of an intricate web of relationships that span across time and space. This insight can guide our ethical decisions, reminding us of the far-reaching impacts of our actions and the inherent value of all parts of the whole.

Fractals embody the principle of eternal becoming, the idea that life is not a static existence but a dynamic process of constant transformation and evolution. This perspective can inspire us to embrace change, to see it not as a threat but as an opportunity for growth and development.

By illuminating these aspects of our existence, fractals provide a powerful lens for interpreting reality. They challenge us to rethink our assumptions, broaden our perspectives, and strive for a deeper understanding of ourselves and the world around us.

6: GENTLE DIP INTO FRACTAL MATHEMATICS

Home is behind, the world ahead,
And there are many paths to tread
Through shadows to the edge of night,
Until the stars are all alight.
Then world behind and home ahead,
We'll wander back and home to bed.
Mist and twilight, cloud and shade,
Away shall fade! Away shall fade!
Fire and lamp, and meat and bread,
And then to bed! And then to bed!
—Peregrin "Pippin" Took in *The Lord of the Rings: The Fellowship of the Ring*, J. R. R. Tolkien

Fractal mathematics is a fascinating field that bridges the gap between abstract concepts and tangible realities, between the finite and the infinite. It is the mathematics of the natural world, the underlying code that patterns the universe, from the smallest particles to the most massive galaxies.

In this part, we will demystify complex mathematical concepts, making them accessible to everyone, regardless of their mathematical background. We will explore the fundamental principles that give birth to fractals, such as recursion, self-similarity, and infinite scaling. We will also introduce the key figures in the history of fractal mathematics, whose innovative ideas and groundbreaking discoveries have shaped our understanding of this domain.

We will also examine the practical applications of fractals

in various fields, such as physics, computer science, and even art. These applications reveal the power of fractals as tools for understanding and shaping our world.

Join us in this fascinating journey as we navigate the landscape of fractal mathematics, a world where art meets science, where the finite intersects with the infinite, and where, as we will discover, dragons lurk in the most unexpected places.

Let's begin our adventure!

Fun with Numbers: Fractals and Number Theory

We begin with a journey into the intriguing world of the Mandelbrot set, a cornerstone of fractal mathematics. As we delve into the complex plane, we unravel the intricacies of the Mandelbrot set and its connection to number theory.

Next, we delve into the fascinating domain of Julia sets, a family of fractal sets intimately related to the Mandelbrot set. Here, we reveal how these sets provide a playground for exploring complex dynamics and their rich interplay with number theory.

We then move on to the topic of Cantor sets, exploring their inherent self-similarity and the role they play in the study of both fractals and number theory. We explore how these sets, though seemingly simple, hide profound complexities within their infinitely nested structure.

The final part of this chapter takes us into the world of the Fibonacci sequence, a sequence synonymous with the golden ratio and the principle of natural growth. We investigate how this sequence manifests itself in fractal form and what it can teach us about the relationship between number theory and fractal geometry.

By the end of this chapter, we hope to reveal how fractal mathematics and number theory are not just isolated disciplines but are intimately linked, each shedding light on the other's mysteries. So buckle up and get ready for a thrilling ride into the world of numbers and fractals!

The Mathematics of the Mandelbrot Set

The Mandelbrot set is perhaps the most famous fractal, and for good reason. Its intricate and beautiful patterns have captivated mathematicians, scientists, and artists alike. Named after the mathematician Benoît Mandelbrot, who made significant contributions to the field of fractal geometry, the Mandelbrot set serves as a remarkable testament to the intricate patterns that simple mathematical rules can create.

At its core, the Mandelbrot set is defined by a deceptively simple iterative process. Given a complex number c, we start with $z = 0$ and repeatedly apply the function $f(z) = z^2 + c$. The complex number c is said to be in the Mandelbrot set if, and only if, the absolute value of z does not go to infinity as we keep applying the function $f(z)$.

Despite the simplicity of its definition, the Mandelbrot set has a remarkably complex structure. The boundary of the Mandelbrot set, in particular, is where the magic happens. Here, we find an infinite variety of shapes and patterns, all of which demonstrate the principle of self-similarity that is characteristic of fractals. Tiny copies of the entire Mandelbrot set appear again and again, each surrounded by a sea of shapes that are similar, but not identical, to the ones that surround the main set.

The Mandelbrot set also has deep connections with number theory. For instance, the periodic points of the Mandelbrot set (points that return to their initial value after a certain number of iterations) correspond to the roots of certain polynomial equations, a central concern in number theory. Moreover, the distribution of these points touches on the field of prime number distribution, further deepening the connection between the Mandelbrot set and number theory.

As we delve further into the mathematics of the Mandelbrot set, we find it to be a rich and fruitful area of study. Not only does it provide striking visual representations of complex dynamics, but it also offers deep insights into the nature of numbers and their relationships. Indeed, the study of the Mandelbrot set provides a fascinating gateway into the world of fractal mathematics and number theory and serves as a testament to the remarkable beauty that lies within the mathematical universe.

The Mandelbrot set is based on complex numbers and the iterative process applied to them, so it's not something that can be directly constructed using simple geometric shapes or lines in the way you might construct a square or a circle. That said, let's look at a conceptual way to understand how it is formed.

1. **Coordinate Plane**: Start with a standard coordinate plane. This gives us the complex plane.
2. **Iterative Process**: Choose a complex number c from this plane. Apply the iterative process $f(z) = z^2 + c$, starting with $z = 0$.
3. **Infinite Iteration**: Continue applying this process an infinite number of times. If the value of z remains bounded (i.e., it doesn't go off to infinity), then the original complex number c you chose is part of the Mandelbrot set.
4. **Coloring**: To visualize this, people usually color points that are in the Mandelbrot set in black, and points that are not in the set in other colors. The color can represent how quickly the values reached infinity, which creates the fascinating and psychedelic images often associated with the Mandelbrot set.
5. **Infinite Complexity**: As you zoom in on the boundary of the set, you will begin to see an infinite amount of intricate detail. You'll find smaller versions of the set, distorted and transformed in various ways, but recognizably the same shape.

It's important to understand that the process of creating the Mandelbrot set is highly dependent on complex numbers and the specific iterative process. While it might be possible to draw a rough approximation of the set or a representation of its general shape, the true set with its infinite complexity and detail can only be fully realized through this mathematical process.

The mathematics of Julia sets begins with understanding complex numbers. Each complex number is composed of two components: a real part and an imaginary part. The imaginary part is a multiple of the square root of -1, which we often denote as *i*. When we square *i*, we get -1. With this understanding, we can express any complex number as $a + bi$, where a and b are real numbers representing the real and imaginary parts, respectively.

The creation of a Julia set isn't a simple one-and-done process, but instead involves repeatedly applying a function in a process called iteration. The function used most frequently in creating Julia sets has a relatively simple form: $f(z) = z^2 + c$. In this equation, z is any complex number, and c is a constant complex number that doesn't change as we iterate the function. The value of c is critical because it dictates the specific shape of the Julia set we're creating.

To determine whether a specific point belongs to the Julia set, we start with a complex number z in the complex plane.

We apply our function to z, generating an output. This output is then fed back into our function as the new z, and the process repeats. We continue iterating in this way, tracking how the values change. If the absolute value of z remains bounded (that is, it doesn't head off to infinity) as we iterate the function an infinite number of times, then the starting point of z is said to belong to the Julia set. If the value of z shoots off toward infinity, then the point is not a part of the Julia set.

In this way, the mathematics of Julia sets encompasses the intricate dance of complex numbers under repeated transformation, a dance that creates the mesmerizing patterns we see in the resulting sets.

Julia sets are a family of mathematical objects related to the Mandelbrot set. While the Mandelbrot set is defined in terms of the behavior of an iterative process as the starting point c varies, a Julia set is defined for a specific value of c.

Here's a basic step-by-step guide for creating a Julia set:

1. **Coordinate Plane**: Like with the Mandelbrot set, we start with a complex plane, with the horizontal axis as real numbers and the vertical axis as imaginary numbers.

2. **Choose a constant**: Choose a complex number c. This will remain constant throughout the process. Each value of c will produce a different Julia set.

3. **Iterative Process**: For each point z in the complex plane, apply the iterative process $f(z) = z^2 + c$. Unlike the Mandelbrot set, this time we are not starting from $z = 0$ but for every point in the plane.

4. **Infinite Iteration**: Apply the process an infinite number of times for each point z. If the value of z remains bounded, then the point is part of the Julia set for that c.

5. **Coloring**: Usually, points in the Julia set are colored black, and points outside the set are colored according to how quickly they diverge to infinity.

6. **Infinite Complexity**: Like the Mandelbrot set, the boundary of the Julia set is infinitely complex. For certain values of c, the set will be connected, forming intricate swirls and spirals. For other values, it will be disconnected or "dusty."

Julia sets are particularly interesting because they offer a way of visualizing the behavior of complex dynamical systems. The intricate patterns they form can provide insights into the chaotic behavior of these systems. Furthermore, the connection between the Mandelbrot set and the Julia sets provides a beautiful illustration of the depth and complexity inherent in the mathematics of fractals.

Let's delve into the Cantor set, another intriguing object in the realm of fractal mathematics. Named after the mathematician Georg Cantor, the Cantor set is a perfect example of a fractal that can be constructed using simple rules, yet it conceals a wealth of intriguing mathematical properties.

The Cantor set is created using a process of iterative removal. We start with a line segment, often represented as the interval from zero to one. This is the first stage of our construction.

In the next stage, we remove the middle third of this segment, leaving two smaller segments: from zero to 1/3 and from 2/3 to one. Now, we have two line segments.

In the third stage, we again remove the middle third from each of these remaining segments. This leaves us with four segments: from zero to 1/9, from 2/9 to 1/3, from 2/3 to 7/9, and from 8/9 to one.

We continue this process indefinitely, removing the middle third from each remaining segment at each stage.

The Cantor set, then, is the set of points that never get removed at any stage of this process. It might seem at first glance that this process would eventually remove all points, but that's not the case. There are points (in fact, an uncountably infinite number of them) that remain through every stage of the process. These points make up the Cantor set.

Remarkably, despite undergoing infinite removals, the Cantor set is not empty. It consists of all points in the original interval [0,1] that can be represented using a ternary (base 3) system where the digits are either zero or two.

But there's more. Although the Cantor set has zero length (since we've removed "all" of the segment), it is uncountably infinite, just like the set of all real numbers between zero and one. This paradoxical result is part of what makes the Cantor set, and fractals in general, so intriguing from a mathematical and philosophical perspective.

Introduction to the Fibonacci Sequence

The Fibonacci sequence is a captivating numerical pattern that has intrigued mathematicians, artists, and scientists for centuries. It is an ordered series of numbers, each being the sum of the two preceding numbers. The sequence begins as follows: 0, 1, 1, 2, 3, 5, 8, 13, 21, 34, 55, 89, 144…and so on indefinitely.

This sequence was introduced to the Western world by Leonardo of Pisa, also known as Fibonacci, in his 1202 book *Liber Abaci*. However, the sequence had been previously described in Indian mathematics. The simplicity of its definition belies the depth and variety of the properties it possesses, making it a subject of continuing interest and study.

To generate the Fibonacci sequence, we start with two

initial numbers, conventionally zero and one. Each subsequent number in the sequence is obtained by adding the two preceding numbers. For example, 0 + 1 = 1, 1 + 1 = 2, 1 + 2 = 3, 2 + 3 = 5, and so on.

One of the unique properties of the Fibonacci sequence is that the ratio of successive terms approaches a particular constant value, known as the golden ratio, as we progress further into the sequence. This ratio, which is approximately equal to 1.6180339887, appears repeatedly in mathematics, art, architecture, and nature.

The Fibonacci sequence, despite its simplicity, is a powerful concept that has far-reaching implications in various fields, including fractal geometry. The recursive nature of the Fibonacci sequence, where the value at each step depends on previous steps, resonates with the recursive nature of fractals, where a simple rule is repeated indefinitely to create complex patterns. In the following sections, we will explore the profound connections between the Fibonacci sequence, the golden ratio, and the world of fractals.

The Golden Ratio and Fibonacci: The golden ratio, often denoted by the Greek letter ф (phi), is a mathematical constant roughly equal to 1.6180339887. It has captivated mathematicians, artists, and architects for centuries due to its unique mathematical properties and its frequent appearance in nature and aesthetics.

But what makes the golden ratio truly fascinating is its intimate connection with the Fibonacci sequence. As we progress further into the Fibonacci sequence, the ratio of two successive Fibonacci numbers (i.e., Fibonacci($n+1$)/Fibonacci(n)) tends to converge to this golden ratio.

To illustrate this, let's look at the first few ratios of successive Fibonacci numbers:

1/1 = 1,
2/1 = 2,
3/2 = 1.5,
5/3 ≈ 1.666,

8/5 = 1.6,
13/8 = 1.625,
21/13 ≈ 1.615…

As you can see, these ratios are getting closer and closer to the golden ratio as we go further in the sequence. By the time we reach the ratio of Fibonacci(144) to Fibonacci(89), we get approximately 1.61803, which is very close to the golden ratio.

This convergence toward the golden ratio is not a coincidence; it is a consequence of the recursive definition of the Fibonacci sequence. This connection becomes clearer when the Fibonacci sequence is expressed in closed form, which involves the golden ratio.

This connection to the golden ratio extends the fascination with the Fibonacci sequence beyond pure number theory. The golden ratio is often associated with aesthetically pleasing proportions in art and architecture, and it also appears in diverse aspects of nature, from the arrangement of leaves on a stem to the spirals of galaxies. In fractal geometry, both the Fibonacci sequence and the golden ratio play crucial roles in the formation of some of the most enchanting fractal patterns.

Fractals and the Golden Ratio: The golden ratio, as we discussed, has a deep connection with the Fibonacci sequence, and through it, with fractal geometry. Many fractals showcase the golden ratio in their structure, embodying this mathematical concept in their infinite, self-similar patterns.

One of the most striking examples is the logarithmic spiral, also known as the growth spiral. This spiral grows outward by a factor of the golden ratio for every quarter turn it makes. That means that if you draw a line from the center of the spiral to any point on it and then draw another line from the center to a point one quarter turn further along, the second line will be approximately 1.618 (the golden ratio) times longer than the first.

This logarithmic spiral isn't just a mathematical abstraction—it can be observed in various aspects of the natural world. For instance, the arrangement of seeds in a sunflower head tends to follow this spiral pattern. As the sunflower grows, each new seed appears at an angle that is the golden ratio of a turn compared to the previous seed. This ensures the most efficient packing of the seeds, allowing the sunflower to fit the maximum number of seeds into a given area.

Another fascinating manifestation of the golden ratio in fractals is seen in the Penrose tiling. This is a method of tiling a plane, discovered by mathematician and physicist Sir Roger Penrose, such that the pattern is nonrepeating but still fills the space completely. The ratios of the areas of the different shapes in the Penrose tiling are related to the golden ratio.

In the realm of fractal geometry, the Fibonacci sequence and the golden ratio serve as critical building blocks in the formation of these intricate and endless patterns. Their presence in the structure of fractals further underscores the pervasive influence of these mathematical concepts in both the natural world and the abstract world of mathematics.

Fibonacci Fractals: Fractals and the Fibonacci sequence are inseparable in many ways, and there are specific fractals directly related to the Fibonacci sequence. These fractals are generated by recursive processes related to the Fibonacci sequence, and they exhibit intricate and fascinating structures that are mesmerizing to explore.

One such fractal is the Fibonacci word fractal. This fractal is created using a Lindenmayer system (or L-system), which is a type of formal grammar most commonly used to model the growth processes of plant development, but it can also be used to generate self-similar fractals.

The process starts with a simple string of characters, say "0." At each step, we apply a certain set of rules to transform the string. In the case of the Fibonacci word fractal, the rule is to replace "0" with "01" and "1" with "0." Starting with "0," the first few steps produce "01," "010," "01001," and so on. This generates a sequence of "words," and these words can be used to create a fractal.

The words are interpreted as instructions for drawing a

line segment: "0" might mean "draw a line segment forward" and "1" might mean "turn left by a certain angle." If you follow these instructions, you'll start to see a fractal pattern emerge. The result is the Fibonacci word fractal, a pattern that shows how the Fibonacci sequence can bridge the gap between numbers and spatial patterns.

Another example of a Fibonacci fractal is the Fibonacci tree. In this fractal, each node at a given level of the tree corresponds to a number in the Fibonacci sequence. The tree grows by adding a new node for each step in the sequence, with the new node sprouting two offspring. This creates a branching structure that mirrors the growth seen in many natural systems, such as the branching of trees or the venation patterns of leaves, and again shows the deep connection between the Fibonacci sequence and fractal structures.

Through these examples, we can see how the Fibonacci sequence provides a numerical foundation for the construction of intricate fractal structures, weaving together the abstract world of numbers and the tangible reality of spatial patterns.

The Fibonacci Sequence in Nature and Fractals: The Fibonacci sequence, despite its simple definition and mathematical elegance, has profound implications for the world around us. It is almost as if nature herself is a mathematician, using this sequence as a blueprint for growth and form. From the intricate spirals of galaxies to the minutiae of a pinecone's scales, the Fibonacci sequence and its associated fractals appear again and again, creating patterns of bewildering complexity and beauty.

One of the most famous examples of the Fibonacci sequence in nature is the arrangement of leaves on a stem, or phyllotaxy. Many plants exhibit a pattern where each leaf is about 0.618 of a turn from the previous one—the golden angle, which is related to the golden ratio derived from the Fibonacci sequence. This arrangement ensures that each leaf gets the maximum amount of sunlight and space to grow, demonstrating how the Fibonacci sequence contributes to the efficiency and survival of the plant.

Similarly, in the world of flora, the branching of trees often follows Fibonacci principles. A tree branch will grow and then split, with each new branch then growing and splitting in turn. This recursive pattern of growth leads to a fractal structure that echoes the recursive nature of the Fibonacci sequence.

Looking closely at pinecones and pineapples, we can observe intricate spirals that trace out Fibonacci numbers. In pinecones, if you count the spirals in one direction, you will typically find a Fibonacci number. Count them in the opposite direction, and you will find an adjacent Fibonacci number. Pineapples display a similar pattern, with distinctive spirals wrapping around the fruit in both directions.

The prevalence of Fibonacci-related patterns in the natural world hints at fundamental principles governing growth and form. It suggests that the Fibonacci sequence and its associated fractals provide a mathematical framework that nature uses to create structures of optimal efficiency, resilience, and beauty. This fascinating interplay between mathematics and nature offers much food for thought, raising profound questions about the underlying mathematical order of the universe and the mysterious ways in which abstract numbers can manifest in tangible reality. The exploration of these questions illuminates the astonishing depth and subtlety of the world around us, inviting us to look closer, delve deeper, and appreciate the mathematical poetry that underlies the tapestry of life.

The Magic of Numbers in Fractals

As we venture deeper into the world of fractals, we uncover an intriguing correlation between these fascinating geometrical figures and numbers. The interplay between fractals and numbers is not just a mathematical curiosity, but a gateway to understanding more profound truths about the universe.

One of the fundamental aspects of this interplay is the way numbers define the structure and behavior of fractals. The Mandelbrot set and Julia sets, which we explored earlier, are quintessential examples of this. The simple iterative process using complex numbers generates the infinitely intricate designs of these fractals.

Similarly, we've seen how the Fibonacci sequence, a simple numerical series, underlies many fractals, with the golden ratio, derived from Fibonacci numbers, appearing repeatedly in the geometric arrangements of numerous natural phenomena. This connection between numbers and fractals extends beyond the Fibonacci sequence. Other number sequences and mathematical functions can also generate fractal structures when applied in a recursive or iterative manner.

The magic of numbers in fractals is not limited to their formative role. Fractals, in turn, have a way of revealing unique properties of numbers. For instance, the Cantor set, a fractal created by repeatedly removing the middle third of a line segment, is a powerful tool for understanding the concept of different sizes of infinity, a core concept in number theory.

Fractals also offer a unique perspective on numbers themselves. In the world of fractals, numbers are not just static values but dynamic entities that can morph and evolve, spinning off intricate patterns and structures through iterative processes. This dynamic view of numbers is radically different from our traditional understanding and hints at a deeper, more fluid reality underlying the rigid numerical framework we are accustomed to.

Moreover, the recursive nature of fractals offers a unique way of understanding the concept of infinity in a concrete and visual manner. The infinite complexity of fractals, arising from simple numerical processes, provides a tangible glimpse into the concept of an actual infinity, something that is otherwise abstract and elusive.

In essence, the magic of numbers in fractals lies in the way they bridge the abstract world of mathematics with the tangible reality of our universe. They serve as a testament to the power of numbers, not just as abstract entities, but as foundational building blocks of our reality. They invite us to view numbers not just as static values, but as dynamic, evolving entities, full of potential for creating intricate patterns and structures. And in doing so, they challenge us to rethink our understanding of the universe and our place within it.

Fractals and Number Theory: A Dance of Complexity and Simplicity
Number theory, an esteemed branch of pure mathematics devoted to the properties and relationships of numbers, unexpectedly intertwines with the world of fractals, the geometric figures defined by self-similarity and intricate patterns. These two seemingly disparate mathematical realms dance together in a ballet of complexity and simplicity, leading to profound insights and a richer understanding of both.

At the heart of this mathematical ballet is the process of iteration, a method of achieving a result by repeatedly applying a function. Both fractals and number theory heavily rely on this process. For instance, the generation of fractals, such as the illustrious Mandelbrot set, involves applying a simple mathematical operation to complex numbers over and over again. This iterative process leads to the creation of infinitely intricate structures from the simplest of mathematical operations. In a similar vein, number theorists often explore the properties of numbers through iterative sequences, shedding light on their intricate structure and behavior.

Further deepening the connection, we find the curious Cantor set, a fractal that elegantly ties into cardinality, a key concept in number theory. Cardinality is a measure of the "size" of a set. The Cantor set, despite appearing deceptively simple, poses intriguing questions about the size and infinity when looked at through the lens of cardinality.

Through the lens of fractals, the seemingly abstract and esoteric concepts in number theory start to take on a tangible form. The complexity of number theory is mirrored in the intricate, self-similar patterns of fractals. In a beautiful symmetry, the simplicity of the mathematical operations used to generate fractals reflects the fundamental nature of the numbers studied in number theory.

So as we delve deeper into the world of fractals, we find ourselves simultaneously exploring the realm of number theory. The two fields dance together, each one leading and following in turn, guiding us on a journey of discovery into the heart of mathematics. It's a testament to the incredible interconnectedness and depth of the mathematical world.

Number theory is a branch of pure mathematics devoted primarily to the study of the integers, or whole numbers. It investigates the intriguing and complex relationships between different types of numbers and the operations that can be performed on them. At its heart, number theory is about exploring patterns, structure, and relationships within the number system.

Fractals, on the other hand, are a concept in mathematical analysis that emerged relatively recently, in the latter half of the twentieth century. They are geometric shapes that are recursively defined, meaning they are created through an iterative process of applying a simple rule or set of rules. Fractals are famous for their self-similarity, which means they appear similar at any level of magnification. In other words, the shape of a fractal appears unchanged, regardless of how much you zoom in or out.

At first glance, these two fields—number theory and fractals—may seem quite disparate. Number theory is ancient, abstract, and concerned with the properties and relationships of numbers, while fractals are a relatively modern mathematical construct that is inherently geometric and spatial in nature. However, as we delve deeper into these topics, we will see that they are more intertwined than they initially appear.

Understanding the relationship between number theory and fractals opens a fascinating gateway to appreciating the harmony and interconnectedness of different branches of mathematics. The bridge between these two domains may initially seem unexpected, but it is through such connections that we can gain a deeper appreciation for the inherent unity of mathematical concepts. This unity not only enhances our understanding of mathematics but also invites us to think more creatively and expansively about the world around us.

Iteration is a fundamental concept that bridges the worlds of number theory and fractals. The word *iteration* comes from the Latin *iterare*, which means "to repeat," and that's precisely what it entails—a process or operation repeated over and over again, often with the output of one stage used as the input for the next.

In the realm of number theory, iteration often leads to the generation of complex sequences and patterns. For instance, consider the simple operation of adding one. If we start with the number one and repeatedly apply this operation, we generate the infinite sequence of natural numbers. This may seem straightforward, but other iterative processes in number theory can yield much more complex and less predictable sequences.

Take, for example, the Collatz conjecture, a tantalizing problem in number theory that remains unsolved to this day. The conjecture involves an iterative process where, given a number, if it is even, you divide it by 2, and if it is odd, you multiply it by 3 and add 1. Despite the simplicity of these rules, when this operation is iterated, it produces sequences of numbers that are incredibly difficult to predict.

Fractals, too, are born out of iteration. A fractal is typically defined by a simple geometric rule or set of rules, which is then applied repeatedly. Each iteration transforms the initial shape, adding layers of complexity. This process continues ad infinitum, each step building upon the last, leading to a final shape that is infinitely complex and detailed.

Consider the famous example of the Mandelbrot set. This set is defined by a simple iterative process involving complex numbers. Despite the simplicity of the rule, when it is applied repeatedly, it generates an image of stunning complexity—a hallmark of fractals.

Through the lens of iteration, we can begin to see the deep connections between number theory and fractals. Both are concerned with the outcomes of repeated processes, and both exhibit a level of complexity that arises from simple starting points. This shared focus on iteration and the complex patterns it generates is a key aspect of the intersection between number theory and fractals.

The Cantor set is a fascinating fractal that lies at the intersection of number theory and geometry, and it provides a tangible representation of some abstract concepts in number theory, such as cardinality.

To begin, let's define the Cantor set. It's created by starting with a line segment (often the interval [0,1] on the real number line) and successively removing the middle third of each remaining segment. After an infinite number of steps, what's left is the Cantor set. At first glance, it might seem like nothing should remain after removing so many segments. However, there are still an infinite number of points left, forming a dustlike fractal set.

This counterintuitive result leads us to the concept of cardinality in number theory. Cardinality is a measure of the "size" of a set, but when dealing with infinite sets, "size" takes on a different meaning than in our everyday experience. The cardinality of a set is not about the physical space it occupies but rather the number of elements it contains.

For example, the set of natural numbers (1, 2, 3...) and the set of integers (...-3, -2, -1, 0, 1, 2, 3...) both have the same cardinality, despite the latter's containing negative numbers as well. This is because there exists a one-to-one correspondence between the two sets—for every natural number, there is a corresponding integer.

Now let's return to the Cantor set. Despite our repeated removal of line segments, the Cantor set still has the same cardinality as the original line segment we started with. This is because there is a one-to-one correspondence between the points in the Cantor set and the points in the original segment. In other words, even though we've removed an infinite number of points, there are still "just as many" points left in the Cantor set.

This tangible example of the Cantor set provides a fascinating insight into the abstract concept of cardinality in number theory. It shows us that in the realm of the infinite, our usual intuition about "size" and "quantity" may not apply. This is just one of the many beautiful connections between fractals and number theory.

Fractals, with their captivating patterns and infinite intricacy, are not just visually appealing—they can also serve as fascinating and tangible representations of abstract mathematical ideas, including those from number theory.

One example of this is the Mandelbrot set, a set of complex numbers defined by a simple iterative algorithm. When visualized, the Mandelbrot set forms a stunning fractal filled with intricate details and recurring patterns. Yet this mesmerizing image is more than just an aesthetic delight. It provides a tangible representation of complex numbers, a concept that can be challenging to grasp intuitively. In the Mandelbrot set, each point represents a complex number, and the color of the point provides information about the behavior of that number when it's repeatedly squared and added to itself. The boundary of the Mandelbrot set, where the colors transition, marks the threshold between numbers that remain bounded under this operation and those that escape to infinity. Hence, the Mandelbrot set is not just a beautiful fractal—it's a visual depiction of a complex mathematical process.

Another example is the Sierpiński triangle, a fractal that starts with an equilateral triangle and recursively removes the middle of each smaller triangle. As this process is repeated infinitely, the Sierpiński triangle remains, a fractal made up of an infinite number of smaller triangles. This fractal is a tangible representation of the concept of infinity, one of the most profound and abstract ideas in number theory and mathematics in general. The Sierpiński triangle illustrates how an object can be infinitely intricate, containing an infinite number of parts, yet remain a finite size.

Both examples highlight the power of fractals as visualizations of number theory. They provide a bridge between the abstract world of mathematical concepts and the tangible world of objects we can see and touch. In this way, fractals not only offer us a new perspective on number theory but also deepen our appreciation for the beauty and complexity of mathematics.

Fractals and number theory might initially seem like disparate fields of study. Fractals, with their stunning complexity and infinite detail, often emerge from simple geometric procedures and rules, while number theory deals with properties and relationships of numbers, particularly integers. However, as we delve deeper into the nature and behavior of these two mathematical domains, we begin to see that they are profoundly interconnected.

One of the most intriguing and meaningful links between fractals and number theory lies in their shared reliance on the concept of iteration. The process of repeating a certain mathematical operation over and over, with the output of one operation becoming the input for the next, is foundational to both domains. For fractals, this iterative process generates intricate patterns and self-similar structures. In number theory, iteration can produce complex sequences and patterns that mathematicians strive to understand.

Consider, for example, the Cantor set, which we explored earlier. This fractal is created through an iterative process of removing the middle third of a line segment. The Cantor set is more than just an interesting geometric figure. It provides a way to visualize and understand the concept of cardinality, an abstract and vital concept in number theory.

Similarly, the Mandelbrot set, which we also discussed, offers a concrete way to visualize and explore the behavior of complex numbers. It's a vivid example of how a simple iterative rule applied to numbers can result in a geometric object of astounding complexity and beauty.

Therefore, the study of fractals can deepen our

understanding of number theory, and conversely, number theory can provide insights into the structure and properties of fractals. The two fields are deeply intertwined, each enriching and informing the other. This unity underscores a fundamental truth about mathematics: its various branches are not isolated islands but part of a vast, interconnected landscape of ideas.

The exploration of the connections between fractals and number theory is not just a fascinating intellectual pursuit. It's also a testament to the unity of mathematics and the amazing depth and intricacy that can emerge from simple rules and patterns. This underscores the beauty, elegance, and profound interconnectedness of mathematical ideas, inspiring us to delve deeper into this captivating world of numbers and patterns.

Complex numbers, a fundamental part of modern mathematics, hold an intriguing place in the world of fractals. Together, they create a rich tapestry of stunning mathematical beauty and offer insights into the nature of numbers and patterns. This chapter will explore the fascinating and intricate relationship between complex numbers and fractals.

Complex numbers expand our numerical system beyond the realm of real numbers by including imaginary numbers. This complex number system is represented as a two-dimensional plane, where the horizontal axis represents the real part of a complex number, and the vertical axis represents the imaginary part.

Meanwhile, fractals, as we know, are intricate geometric structures exhibiting self-similarity and formed through iterative processes. What's fascinating is that some of the most well-known and visually stunning fractals, such as the Mandelbrot set and Julia sets, are created using complex numbers.

The iterative processes that generate these fractals take complex numbers as inputs and outputs. When you plot the results on the complex plane, you get figures of astonishing complexity and beauty. The Mandelbrot set, for instance, is a set of complex numbers that remains bounded under repeated iteration of a certain function. It's remarkable that such a simple rule can generate such a complex and infinitely detailed structure.

We will delve deeper into the world of complex numbers and their relationship with fractals. We will explore how complex numbers are used to generate fractals, and how fractals, in turn, provide visual representations of complex numbers and their behaviors.

Moreover, we will explore the philosophical implications of this relationship, reflecting on what it tells us about the nature of numbers, patterns, and the mathematical universe. Through this journey, we will gain a deeper appreciation of the intricate beauty and profound interconnectedness of mathematical ideas.

Introduction to Complex Numbers

Complex numbers are a unique and fascinating type of number that exist beyond the familiar realm of integers, fractions, and real numbers. Unlike the numbers most people encounter in everyday life, complex numbers are made up of two parts: a real part and an imaginary part. This is typically expressed in the form $a + bi$, where a and b are real numbers, and i is the square root of -1, a number that doesn't have a real value and is thus referred to as imaginary.

In the world of complex numbers, i is a fundamental entity, and it plays a crucial role in the arithmetic operations of complex numbers. Addition, subtraction, multiplication, and division all have special rules when dealing with complex numbers, rules that may at first seem counterintuitive but soon reveal a symmetrical and harmonious mathematical structure.

These numbers are plotted on a plane known as the complex plane, which is similar to the Cartesian plane used for graphing real numbers. In the complex plane, the horizontal axis represents the real part of the number, while the vertical axis represents the imaginary part. This way, every point on the complex plane corresponds to a unique complex number.

Although the concept of complex numbers may seem abstract and perhaps even absurd, they play a crucial role in many areas of mathematics and physics, including the study of fractals. Understanding complex numbers, their properties, and their behavior under various mathematical operations is the key to unlocking the intricacies of fractal geometry.

Complex Numbers and Fractals: The connection between complex numbers and fractals is profound and intrinsic. The most famous fractal, the Mandelbrot set, is defined by a simple equation involving complex numbers:

$z = z^2 + c$

where both z and c are complex numbers.

This equation, although seemingly simple, leads to a world of complexity and infinite detail.

To create the Mandelbrot set, we start with a complex number for c and set z to zero. Then we repeatedly apply the equation, each time using the result as the new z. This process is known as iteration. Depending on the behavior of z as we keep iterating, we can decide whether the initial complex number c belongs to the Mandelbrot set or not.

If the value of z remains bounded (that is, it doesn't go off to infinity) after infinite iterations, then the original complex number c is part of the Mandelbrot set. If z spirals off to infinity, then c is not part of the set. This simple rule leads to an astonishingly complex structure when visualized on the complex plane.

Similarly, Julia sets, another class of fractals, are also defined using complex numbers and iteration. Each Julia set is associated with a unique complex number, and its structure can be as intricate and fascinating as that of the Mandelbrot set.

Through these and other examples, it becomes apparent that complex numbers are not just an abstract mathematical concept, but a powerful tool for understanding and generating the intricate and beautiful structures of fractals.

Visualizing Complex Numbers through Fractals: One of the challenges with complex numbers lies in their visual representation. Since they're composed of two parts (real and imaginary), they cannot be easily represented on a one-dimensional number line like real numbers. This is where the complex plane comes in, allowing us to visualize complex numbers in two dimensions, with the real part on the horizontal axis and the imaginary part on the vertical axis.

Fractals, particularly the Mandelbrot and Julia sets, offer a compelling way to visualize complex numbers and their behaviors. By assigning each point in the complex plane to a color based on whether the corresponding complex number belongs to the set, we can create vivid and intricate images that provide a visual representation of the set.

For example, in the Mandelbrot set, the color of each point represents the rate at which the corresponding complex number goes off to infinity under the iteration of the function $z = z^2 + c$. This results in a stunning image full of elaborate detail and infinite complexity, all derived from the behavior of complex numbers.

The Julia sets, on the other hand, allow us to visualize the behavior of a single complex number when iterated under the same function. Each Julia set corresponds to a unique complex number, and the resulting images can range from simple and symmetric to wildly intricate, depending on the chosen number.

Through such visualizations, fractals not only make the abstract concept of complex numbers more tangible, but they also reveal the inherent beauty and complexity within this area of mathematics. They show us that every point in the complex plane has a story to tell, full of unexpected twists and turns, a story that can be as complex and fascinating as the numbers themselves.

The Behavior of Complex Numbers in Fractals: When we examine the role of complex numbers in fractal generation, particularly in the creation of the Mandelbrot and Julia sets, we embark on a journey that illuminates the beautiful, unpredictable, and often wild behavior of these numbers.

In the context of the Mandelbrot set, each point on the

complex plane corresponds to a particular complex number. We can take each of these numbers through an iterative process, feeding it into a simple quadratic equation:

$z = z^2 + c$.

Here, z starts as 0 and c is the complex number corresponding to our current point in the plane. The result is then fed back into the equation as the new z, and the process repeats.

If the absolute value of z remains bounded (that is, it doesn't tend toward infinity) after many iterations, then the complex number c is part of the Mandelbrot set. The intricacy of the set comes from the points that are just on the edge—where the behavior of the numbers becomes chaotic, neither clearly diverging to infinity nor staying completely bounded.

For Julia sets, the process is similar, but with a key difference: the value of c is held constant for all points in the plane, and the value of z starts as the complex number corresponding to each point. Different values of c yield dramatically different Julia sets, some showing beautiful symmetry, others exhibiting a chaotic, infinitely intricate structure.

Understanding the behavior of complex numbers in this context is not just about knowing if they will eventually escape to infinity or remain bounded. The true magic lies in the dance they perform on the edge of chaos, where the slightest change can lead to dramatically different outcomes. This is where fractals, and the complex numbers that generate them, truly come to life—on the boundary between the known and the unknown, the ordered and the chaotic, the finite and the infinite.

Complex Numbers, Fractals, and Real-World Applications: Understanding the relationship between complex numbers and fractals is not just an esoteric exercise for mathematicians. This relationship has real-world applications that impact various fields, from physics to computer graphics and even biology.

In physics, fractals and complex numbers are used in

the study of chaotic systems. These systems, such as weather patterns or fluid dynamics, exhibit behavior that is highly sensitive to initial conditions. By using fractals and complex numbers, scientists can model these systems more accurately and gain a better understanding of their underlying dynamics.

In the field of computer graphics, the visualization of fractals generated from complex numbers has led to the creation of stunningly beautiful and infinitely detailed images. These images are not just aesthetically pleasing but are also used in creating realistic computer-generated landscapes and special effects in movies.

In biology, the concept of fractals has been applied to model various natural phenomena. The structure of certain biological systems, like blood vessels or the branching pattern of lungs and trees, exhibit fractal-like patterns. By using complex numbers and fractals, biologists can simulate these structures more accurately, leading to a better understanding of their form and function.

The exploration of complex numbers and their relationship with fractals reveals a world where art and science, aesthetics and computation, chaos and order intersect. This intersection is not just theoretical; it has practical implications that enhance our understanding of the world and enable us to simulate, create, and explore with greater depth and realism. Thus, the fascinating world of complex numbers and fractals is not just a mathematical curiosity but a tool that enhances our capacity to understand and navigate our complex world.

7: PLAYING WITH PATTERNS: ALGORITHMS AND FRACTALS

Mathematical reasoning may be regarded rather schematically as the exercise of a combination of two facilities, which we may call intuition and ingenuity.
—Alan Turing

In our journey to understand the fascinating world of fractals, we have touched upon their manifestations in nature, art, philosophy, and mathematics. In this chapter, we venture into the realm of algorithms, the systematic procedures or formulas for solving problems, and their fundamental role in the creation and understanding of fractals.

Algorithms are everywhere in our digital world. They form the backbone of our computer systems, enabling us to perform complex tasks, make predictions, and automate processes. But what if we told you that these same algorithms are key to unlocking the mysteries of the fractal universe?

In this chapter, we will explore the intimate relationship between algorithms and fractals, looking at how simple iterative processes can give rise to complex and beautiful patterns. We will delve into the algorithms that generate famous fractals like the Mandelbrot set, the Sierpiński triangle, and the Koch snowflake, and examine how these procedures mirror the iterative processes found in nature.

So prepare to step into the fascinating interplay of structure and chaos, determinism and unpredictability, as we decode the algorithms of the fractal universe. Hold on tight, for it's going to be a thrilling ride into the mathematical heart of nature's intricate patterns.

Decoding Patterns: What Are Algorithms?

Algorithms are the hidden orchestrators of our digital world. They are essentially step-by-step instructions that guide us through tasks, similar to a recipe. Just as a recipe provides a methodical process to transform individual ingredients into a delicious dish, an algorithm offers a systematic procedure to take an input, process it, and produce a desired output.

While we often associate algorithms with computer science and programming, they've been part of human problem solving long before the advent of computers. For example, the ancient Egyptians used an algorithmic approach to solve multiplication problems, and Euclid's algorithm, developed around 300 BCE, efficiently calculates the greatest common divisor of two numbers.

Algorithms can be simple or incredibly complex, depending on the problem they're designed to solve. Importantly, every algorithm has a specific purpose, whether it's sorting a list of numbers, finding the shortest route from point A to point B, or generating a fractal image.

In the context of fractals, algorithms play a fundamental role. Fractals are often defined and generated through iterative algorithms, which take an initial input, transform it according to specific rules, and then use the output as the new input for the next iteration. This process is repeated indefinitely, creating intricate patterns that are the hallmark of fractals.

This interplay between simplicity and complexity, order and chaos, is a defining feature of fractals and a testament to the power of algorithms. By following a set of simple instructions repeatedly, we can produce structures of stunning intricacy and beauty, echoing the processes that shape the patterns of our natural world.

In the following sections, we will delve deeper into the world of algorithms and their role in creating some of the most famous fractals. We will see how these seemingly abstract mathematical constructs provide a bridge between the world of numbers and the tangible, visible universe around us.

Iterative Processes in Algorithms: Iteration, a fundamental concept in both computer science and fractal geometry, is the process of repeating a set of instructions. In algorithms, iteration is the procedure of executing loops—a series of steps that are performed repeatedly until a certain condition is met. This mechanism of repeated execution allows us to handle vast amounts of data and perform complex tasks efficiently.

But how does this notion of iteration connect to fractals? It is through iteration that fractals take shape. Fractals, as we have seen, are self-similar, meaning they display the same pattern at various scales. This self-similarity is achieved through repeating a simple process over and over again, a concept known as recursion.

Recursion in fractals works in a similar way to iteration in algorithms. For instance, consider the Sierpiński triangle, a well-known fractal. To create a Sierpiński triangle, you start with a single equilateral triangle. Then, you remove the triangle formed by connecting the midpoints of its sides, resulting in three smaller equilateral triangles. This process is then repeated for each of the remaining triangles, ad infinitum. The iterative, recursive nature of this process is what gives rise to the fractal's complex, self-similar structure.

In the realm of computer graphics, these iterative processes are encoded into algorithms to generate intricate fractal images. These algorithms run loops that apply a simple transformation repeatedly to every point in an image, creating the repeating, infinitely detailed patterns that characterize fractals.

In this way, iteration serves as a bridge between the worlds of algorithms and fractals. Through the mechanism of iteration, simple rules and processes can give rise to the complex, self-similar structures that are the hallmark of fractal geometry.

Algorithms and the Mandelbrot Set: The Mandelbrot set, arguably the most famous fractal, is the perfect illustration of how algorithms and fractals intertwine. This complex and infinitely detailed object is generated by a simple algorithm, one that iteratively applies a mathematical function to complex numbers.

The algorithm begins with a complex number c and a starting number usually set to zero. In each iteration of the algorithm, the current number is squared and c is added to the result. The algorithm repeats this process for a set number of iterations or until the absolute value of the current number exceeds a certain threshold.

The key to the Mandelbrot set is in determining for which complex numbers c the sequence remains bounded. If the sequence remains bounded within a certain range, the complex number c is said to be in the Mandelbrot set and is usually colored black on images of the set. If the sequence goes beyond the threshold, the number c is not in the Mandelbrot set, and it is usually colored according to how quickly the sequence exceeded the threshold.

Despite its simplicity, when this algorithm is applied millions or even billions of times, it generates an object of extraordinary complexity and beauty. The boundary of the Mandelbrot set, in particular, displays an intricate, infinitely repeating pattern that exemplifies the essence of fractal geometry. It is the perfect example of how simple algorithms can create complex and beautiful fractals.

Algorithms and the Koch Snowflake: Another classic example of the connection between fractals and algorithms is the Koch snowflake, a mathematical curve and one of the earliest fractal curves to have been described.

The algorithm to generate the Koch snowflake is quite straightforward. We start with an equilateral triangle, which will be our initiator. The generator, or the shape that is to be repeated, is a smaller equilateral triangle that we use to replace the middle third of each side of our initiator.

For each side of the equilateral triangle:
1. Divide the line segment into three equal parts.
2. Create an equilateral triangle with the middle segment as the base, pointing outward.
3. Remove the line segment that was the base of the new equilateral triangle.

This process is then repeated indefinitely with the new shape. Each line segment on the shape undergoes the same process, and with each iteration, the outline of the snowflake becomes increasingly complex, creating a beautiful fractal pattern.

The Koch snowflake is notable for having an infinite perimeter but a finite area. This property demonstrates one of the paradoxical aspects of fractals, showing how they can challenge our conventional understanding of dimensions and geometry. It is through the simple algorithmic process described above that such fascinating properties can arise.

Algorithms and the Sierpiński Triangle: The Sierpiński triangle, named after the Polish mathematician Wacław Sierpiński, is another fractal that can be generated through a simple algorithm.

The algorithm to generate a Sierpiński triangle is as follows:
1. Start with an equilateral triangle. This is our initiator.
2. Divide this triangle into four smaller equilateral triangles by connecting the midpoints of each side.
3. Remove the middle triangle, leaving three equilateral triangles.
4. Repeat steps 2 and 3 for each of the remaining smaller triangles indefinitely.

Each iteration of the algorithm results in a shape with

more triangles removed, creating a pattern of increasingly complex holes within holes.

The Sierpiński triangle is a great example of how simple rules can produce intricate patterns. It illustrates the concept of self-similarity, as each smaller triangle is a scaled-down copy of the larger triangle. This reflects one of the key properties of fractals—no matter how much you zoom in or out, the pattern remains the same.

The Sierpiński triangle also has fascinating mathematical properties. For example, despite the infinite number of triangles it contains, the total area of all the triangles is finite, while the perimeter is infinite.

This simple algorithm, starting with a single shape and applying a rule repeatedly, illustrates the power of iterative processes in creating fractal structures.

Fractal Algorithms and Nature: Fractals are not only intriguing mathematical constructs but are also fundamental to the structure of the world around us. Nature is replete with examples of fractals, from the branching of trees and rivers to the structure of clouds and mountains. And at the heart of these patterns lie algorithms.

Consider the growth of a tree. It starts as a single sprout, which grows and branches out. Each branch then grows and further branches out, and this process repeats over and over. This growth can be thought of as a simple algorithm: "grow, then branch."

Similarly, the formation of a river system follows a kind of algorithm. Water flows downhill, eroding the landscape as it goes, creating small rivulets that join to form larger streams, which combine to form even larger rivers. The pattern of a river system, with its intricate network of tributaries, is a fractal, and it's generated by the repeated application of the "flow and join" algorithm.

These examples illustrate how fractal algorithms are not just abstract mathematical concepts but are deeply embedded in the processes that shape our natural world. Understanding these algorithms can give us insights into the patterns of nature and how complex structures can emerge from simple rules.

Applications of Fractal Algorithms: Fractal algorithms have far-reaching applications that extend beyond the natural world. In the realm of technology, they play a vital role in data compression, image rendering, and signal processing. The concept of fractals is foundational in the field of computer graphics where intricate, realistic patterns and images are created using simple mathematical rules.

For instance, in the film industry, fractal algorithms are used to create realistic animations of natural phenomena such as fire, smoke, water, and terrain. The reason these simulations look so authentic is that they mirror the same kind of complexity and self-similarity found in real-world systems.

In the realm of data compression, fractal algorithms are used to store large amounts of information in a smaller space. This is done by identifying and encoding repeating patterns within the data. The result is an efficient way of storing complex data that can be accurately reconstructed when needed.

Moreover, in the field of medicine, fractal geometry has been used to analyze and predict patterns in everything from DNA sequences to heart rates to the growth of cancer cells, providing new insights and advancements in healthcare.

Thus, the application of fractal algorithms is vast and continually evolving, demonstrating their integral role in our technologically advanced society.

Fractal Algorithms and Art: Art, whether traditional or digital, has been greatly influenced by fractal algorithms. This is perhaps most evident in the field of digital art, where fractal algorithms have been used to create stunning images that capture the imagination with their intricate patterns and almost infinite levels of detail.

In traditional art, fractal patterns have been observed in the works of artists long before the term "fractal" was coined. From the recursive patterns in Romanesco broccoli depicted in Renaissance paintings to the intricate designs in Islamic art, the influence of fractal-like patterns is not a new phenomenon.

In the digital era, artists have harnessed the power of fractal algorithms to create mesmerizing works of art. Digital art software often uses fractal algorithms to generate images that are infinitely complex, yet they are formed by the repeated application of simple rules. This allows artists to explore the aesthetic potential of mathematical patterns in a way that would be impossible with traditional mediums.

Computer-generated fractal art often has an otherworldly quality, with intricate spirals, swirls, and feathers of color. These works can evoke a sense of the infinite, as the same patterns repeat at different scales, inviting the viewer to explore the image in ever-greater detail.

Whether in the digital realm or the canvas, fractal algorithms have provided artists with a new toolset for expressing their creativity, bridging the gap between art and mathematics.

Fractal Algorithms in Computer Science: Fractal algorithms have significant applications in the field of computer science, particularly in data compression and graphics rendering. In data compression, fractal algorithms are used to reduce the amount of data needed to represent images and other types of information. By identifying and encoding repeating patterns at different scales, fractal compression can achieve high compression ratios, often with minimal loss of quality. This is particularly useful for applications such as satellite imagery and medical imaging, where large amounts of data need to be transmitted or stored efficiently.

In graphics rendering, fractal algorithms are used to create complex, realistic images from simple rules. For example, in computer graphics and video games, fractal algorithms can generate realistic terrain by repeating patterns at different scales, creating mountains, valleys, and other natural features with a high level of detail. Similarly, fractal algorithms can create realistic textures for surfaces, like the bark of a tree or the surface of water, by repeating patterns at different scales.

Beyond these applications, the principles of fractal algorithms have also influenced the design of data structures and algorithms in computer science. For instance, trees and graphs, which are fundamental data structures in computer science, have fractal-like properties. Many important algorithms, like quicksort and merge sort, have a recursive structure similar to fractals.

By harnessing the power of repeated patterns and recursion, fractal algorithms have proven to be a powerful tool in computer science, enabling us to represent, process, and generate complex information efficiently and effectively.

Fractals, Algorithms, and the Universe: A Reflection: In this final section, we return to the broader themes of our exploration. Fractals and algorithms are more than mathematical curiosities or practical tools. They offer profound insights into the nature of our universe and our place within it.

Fractals, with their self-similarity across scales, remind us of the interconnectedness of all things. From the spirals of galaxies to the spirals of DNA, the same patterns repeat at every level of existence. This reveals a deep unity underlying the apparent diversity and complexity of our universe.

Algorithms, on the other hand, show us the power of simple rules to generate complexity. From the basic operations of arithmetic to the rules of cellular automata, complexity emerges not from chaos but from order and regularity. This suggests that the complexity of our universe may not be random or accidental but the product of simple, underlying laws.

Together, fractals and algorithms suggest a view of the universe as a complex, interconnected system, generated by simple rules and repeating patterns. They remind us that we are not separate from the universe but part of it, embedded in its patterns and subject to its laws. As we continue to explore the mysteries of our universe, fractals and algorithms will no doubt continue to guide us, offering new insights and new perspectives. They are not only tools for understanding the universe, but also metaphors for our place within it, reminding us of the beauty, complexity, and interconnectedness of all existence.

8: FRACTALS IN ECONOMICS TO EVOLUTION

Chaos theory throws it right out the window. It says that you can never predict certain phenomena at all. You can never predict the weather more than a few days away. All the money that they put into meteorology, all the decades of study, have not extended the range of prediction at all. And it never will. There is a kind of built-in level of unpredictability.
—Ian Malcom in Michael Crichton's *Jurassic Park*

As we delve into the world of economics, we find ourselves standing at the intersection of human behavior, society, and mathematics. Amid the flux of supply and demand, inflation rates, and fiscal policies, there's a hidden pattern, a secret language that offers us a fresh perspective on this complex discipline. That secret language is spoken in fractals.

Fractals, these intricate patterns repeating ad infinitum, might seem too abstract or too mathematical to have any relevance in the bustling, pragmatic world of economics. However, the story of fractals in economics is not unlike finding a hidden key in a dusty old book. It opens a way to view economic data and trends not as separate, isolated entities but as interconnected parts of a larger, complex whole.

The genesis of this idea takes us back to the pioneering work of Benoît Mandelbrot, who applied fractal geometry to economic data. Mandelbrot's innovative thinking nudged economists to consider markets as "wild" rather than "mild," fraught with abrupt changes and unpredictable variations. This radical perspective may not have turned economics on its head overnight, but it did plant the seed for a new approach to understanding financial markets.

In the world of finance and economics, trends and data often seem erratic and haphazard. However, just like a Jackson Pollock painting, what at first glance appears chaotic, upon closer inspection, reveals an underlying structure and pattern—a fractal nature. By applying the principles of fractals, analysts can better understand the wild and complex nature of financial markets.

And so we set off on our journey to explore how the mysterious world of fractals illuminates the twisting, turning paths of economics. From the highs and lows of stock market trends to the complex dance of interest rates and inflation, we'll see how fractals provide a new language to narrate the grand story of economics. Brace yourselves; it's going to be an exciting ride!

Now that we've established our setting at the crossroads of fractals and economics, let's stroll down the first path of our journey—the fractal nature of financial markets.

Imagine, if you will, a day at the stock exchange. The incessant ringing of phones, the frantic shouts of brokers, the flickering numbers on giant screens: it's a world that thrives on chaos, an unpredictable dance of supply and demand. But as we've already hinted, there's a rhythm beneath this seeming randomness, a melody composed in the language of fractals.

You see, markets have a fascinating trait. They're self-similar, meaning patterns that occur over shorter periods, like days or weeks, mirror those happening over longer spans, like months or years. This trait, which is a hallmark of fractals, suggests that financial markets aren't as disordered as they first appear. They possess an inherent structure, albeit a complex one.

Take, for instance, the price movements of stocks or commodities. You might have heard of "bubbles" and "crashes." They're periods of rapid price increase followed by a sharp fall, and they seem to recur with astonishing regularity. These fluctuations might seem random, but viewed through the lens of fractal geometry, they reveal self-similar patterns.

But how do fractals help us here, you might wonder? For one, they provide a tool to analyze and understand these market movements better. Fractal analysis can help us identify underlying trends, make more accurate predictions, and even craft more robust financial models.

So amid the cacophony of the stock exchange, fractals offer a way to make sense of the madness. It's as if we're listening to a complex symphony, and suddenly we start recognizing the repeating themes and variations. That's the power of fractals—they help us see the order in the chaos of financial markets.

In section three we'll dive deeper into how the principles of fractal geometry apply to market risk assessment.

Risk is an inherent part of any economic activity. From the individual deciding where to invest their hard-earned savings to the multinational corporation planning its next big move, everyone grapples with uncertainty. Traditional risk assessment methods typically rely on standard statistical methods, such as Gaussian distribution. However, these models have their limitations. They often underestimate the likelihood of rare events—the so-called "black swans."

Enter fractal analysis, with its power to describe irregular and complex systems. By looking at market data through the lens of fractal geometry, we can discern patterns that conventional models might miss. Fractal models can account for sudden, drastic market shifts—the black swans—by acknowledging the market's inherent unpredictability and complexity.

Fractals help us realize that the world of finance is more akin to a turbulent ocean than a placid lake. It's an environment of constant flux, where calm periods could suddenly give way to turbulent storms. Traditional models often fail to account for these abrupt changes, leading to catastrophic miscalculations.

By contrast, a fractal-based approach assumes such turbulence is not just possible, but inevitable. It helps us prepare for drastic market movements rather than being caught off guard. In this sense, fractal analysis can equip traders, investors, and economists with a more realistic, robust tool for assessing market risks.

In short, the role of fractals in financial risk assessment can be likened to a lighthouse, providing vital guidance amid the turbulent, unpredictable sea of economic activity. And just like any seasoned sailor will tell you, recognizing the true nature of the sea—its unpredictability—is the first step to navigating it successfully.

The fourth section of our journey in this economic landscape brings us to the intriguing concept of the fractal markets hypothesis. This idea was proposed as a counterpoint to the efficient market hypothesis, which asserts that it is impossible to beat the market since stock market efficiency causes existing share prices to always incorporate and reflect all relevant information.

The efficient market hypothesis presumes that the distribution of stock price changes is normal and that these changes are independent of each other. In layman's terms, it suggests that market movements are essentially a random walk, devoid of any discernible pattern.

However, the fractal markets hypothesis begs to differ. It proposes that markets are far from random and that they exhibit a specific kind of order—a fractal order. Much like how a fern's shape is repeatedly replicated in each of its leaves, market movements at different scales can often mirror each other.

To understand this, imagine zooming into a graph of stock prices. You might see large, sharp peaks and troughs representing days with drastic price changes. Now if you were to zoom out and look at the graph over a decade, you would see a similar pattern of peaks and valleys, albeit at a much larger scale. This self-similarity, irrespective of the scale, is a hallmark of fractal systems.

The fractal markets hypothesis also posits that market conditions are not always the same. There are periods of high volatility and periods of stability, and these can be tied to the fractal nature of the market. By acknowledging and understanding these fractal properties, traders and economists can get a better grasp on market behavior.

In this light, the fractal markets hypothesis is not just a theoretical concept; it's a more nuanced lens to understand market dynamics, offering new avenues to navigate the intricate pathways of economic exchange.

In section five we will delve into the realm of chaos theory and its fascinating intersections with economics. To understand chaos theory in the context of economics, we need to travel back in time to the discovery of the butterfly effect by Edward Lorenz.

While working on a mathematical model for weather prediction, Lorenz stumbled upon an intriguing phenomenon. He found that tiny differences in initial conditions led to vastly different outcomes in his model over time. A change as minuscule as the flap of a butterfly's wings in Brazil, he suggested, could set off a tornado in Texas—hence the name "butterfly effect."

This concept is central to chaos theory, which describes how small changes in a complex system can lead to dramatic variations in future states of that system. Now if you think about it, an economy is a quintessential example of a complex system. It's composed of numerous individual elements —businesses, consumers, government bodies, financial institutions—all interacting and influencing each other in myriad ways.

So how does chaos theory and the butterfly effect apply to economics and fractals? Let's take the stock market as an example. Suppose a minor piece of news about a small tech startup gets leaked. It may initially affect only the startup's stock price. However, this fluctuation could influence investors' perceptions about the tech sector, leading to a cascade of buying or selling of other tech stocks. Over time, these effects could amplify, causing significant shifts in the overall stock market.

This, in turn, reflects the fractal nature of the market, as minor changes in one part of the market (a single stock) can impact the larger system (the overall market), creating patterns that are similar across different scales.

In this sense, chaos theory and the concept of the butterfly effect help us appreciate the complexity and unpredictability of economic systems. They also offer us a different perspective, one that recognizes the inherent instability and dynamism of these systems, providing yet another reason to be cautious about oversimplifying economic behaviors and trends.

Section six will explore the concept of power laws in the context of economics and how they relate to fractals. Power laws are a type of mathematical relationship between two quantities, where a change in one quantity leads to a proportional relative change in the other quantity.

For example, in the distribution of wealth, power laws dictate that the percentage of people who have a certain wealth is inversely proportional to a power of the wealth itself. This pattern, observed in many countries, shows an unequal wealth distribution, where the majority of wealth is held by a small percentage of the population.

This distribution can be visualized as a fractal, where each segment of the distribution (representing a group of people with a specific wealth range) looks similar to the entire distribution. This is because fractals are, by definition, self-similar; the same pattern repeats at all levels of scale. This principle helps economists visualize complex and vast data sets in a more comprehensible form.

Moreover, understanding power laws can help economists model economic phenomena more accurately. For instance, power laws can provide insights into market crashes. It's observed that large changes in stock prices, though less frequent, follow a similar distribution as smaller changes, a pattern that can be modeled by power laws.

Recognizing this power-law relationship could lead to better predictive models for economic behavior. However, just like the weather patterns Edward Lorenz was trying to predict, economies are complex systems subject to the butterfly effect. While power laws and fractals provide us with tools to understand and potentially predict economic patterns, they also reveal the inherent uncertainty and unpredictability of economic systems.

Hence, fractal mathematics offers not only a way to model and understand economic systems, but also a philosophical viewpoint, a shift in our understanding and interpretation of economic behaviors and realities. It's about recognizing the patterns while also acknowledging the complexity, the chaos, and the wonder that lies therein.

Risk assessment is a crucial part of economics, especially in areas such as insurance, finance, and investments. Traditionally, risk assessment models relied on standard probability distributions, such as the Gaussian (bell-curve) distribution. These models assumed that all possible outcomes and their probabilities could be precisely defined, and large deviations from the average were extremely rare.

However, real-world financial data often show a different story—large changes or "black swan" events are more common than traditional models would suggest. The 2008 financial crisis is a stark reminder of such outliers that could cause catastrophic outcomes.

Here is where fractals enter the stage. Benoît Mandelbrot, the father of fractal geometry, pointed out that financial data often exhibit "heavy tails," meaning that extreme events are more likely than Gaussian distributions predict. This characteristic can be modeled with fractals, providing a more realistic representation of financial markets' behavior.

Take, for instance, a coastline. If we measured it with a large stick, we would get a particular length. However, if we used a smaller stick, we'd measure more details and get a longer length. This self-similar property, where the amount of detail increases with finer measurements, is a key characteristic of fractals, and it's also seen in financial data. If we look at stock market data over a year, we might see a certain amount of variation. But if we look at daily or hourly changes, we see a lot more volatility—the "roughness" increases, just like in fractals.

By applying fractal mathematics to risk assessment, economists and financial analysts can design models that are more aligned with real-world market behavior, accounting for the unexpected extreme events and their significant impacts. However, these models also underscore that, much like weather systems, financial markets are intrinsically complex and unpredictable. Predicting a black swan event is still largely beyond our reach, but recognizing that they are part of the system's inherent behavior is a step forward in understanding risk in economic systems.

Just as networks of neurons create the complexity of human thought, economic networks generate the complexity of economic systems. These networks—comprised of connections between firms, markets, countries, and even individual consumers—exhibit patterns that are neither entirely ordered nor entirely random.

Enter fractals. The self-similarity property of fractals provides an intuitive model for understanding these complex networks. For instance, the network of a global corporation might look very similar at different scales: the CEO at the center, connected to senior executives, who are in turn connected to middle managers, and so forth, all the way down to individual employees. Each level of the hierarchy is a smaller-scale copy of the one above it.

What's more, these networks often show scale-free properties: just as a fractal looks the same at any scale, the distribution of connections in these networks remains the same no matter how much you zoom in or out. This feature can be seen in many economic networks, from connections between banks to patterns of trade between nations.

The fractal nature of economic networks has significant implications. One key takeaway is their robustness. Fractal networks can lose any one node without the network collapsing, which speaks to the resilience of many economic systems. However, they're also vulnerable to coordinated attacks—if too many centrally connected nodes are removed, the entire network can disintegrate.

Moreover, the insights gained from studying fractal economic networks can guide policymakers in designing more robust economic systems, identify points of vulnerability, and even anticipate the spread of financial crises. For instance, understanding the network structure could have helped regulators foresee the cascading failures of financial institutions during the 2008 economic crisis.

Like a fractal, the complexities of economic systems emerge from simple rules applied repeatedly, resulting in intricate networks. Understanding these networks can help us not just understand our economy better, but also guide its future in a more informed way.

Fractals aren't just a tool to understand the current structure of economic systems; they also provide a means for predicting future economic patterns. The idea of predictive modeling in economics is not new—economists have long used various models to predict outcomes based on various factors. But fractal mathematics offers a new dimension to these predictive models, as it accounts for the inherently complex, nonlinear nature of economic systems.

Recall the underlying principle of fractals: simple rules iteratively applied yield intricate patterns. This concept maps perfectly onto many economic phenomena, where simple interactions between individual market participants give rise to complex market behaviors.

A great example of this is in the field of financial markets. It is well known that financial markets exhibit "fat tails"— the idea that events that are highly unlikely under a normal distribution occur more frequently than expected. This can be modeled using fractal mathematics, leading to more accurate risk assessments.

On a larger scale, fractal patterns can be seen in the cycles of boom and bust that characterize economic systems and in the patterns of innovation and technological change. Traditional linear models may miss these complex patterns, leading to forecasts that underestimate the likelihood of extreme events, such as economic recessions or rapid technological advances.

By using fractal models, economists can create more accurate predictive models that account for these complexities. Of course, as with all models, fractal models can't predict the future with 100 percent accuracy. But by providing a closer approximation of the intricacies of economic behavior, they can help us make more informed decisions, potentially helping to mitigate the effects of future economic downturns or to better harness the opportunities presented by technological change.

In this context, fractal mathematics can be seen as another tool in the economist's toolbox—a tool that, when used wisely, can provide valuable insights into the patterns underlying the economy.

As we have seen throughout this chapter, fractal mathematics has a great deal to offer when it comes to understanding, analyzing, and predicting economic phenomena. This insight has opened the door to new ways of thinking about the economy, allowing us to better comprehend the complexities that drive market behavior and economic cycles.

In the future, the use of fractal thinking in economics may lead to the development of new policies and strategies that are better equipped to manage the inherent complexities and nonlinearities of the economic systems. Economists and policymakers who embrace the lessons of fractal mathematics are likely to have a more nuanced understanding of the factors that influence economic performance, enabling them to make better decisions for both short-term and long-term growth.

Furthermore, the interdisciplinary nature of fractal mathematics means that its principles can be applied to a wide range of fields. Economists who take a fractal approach to their work may find themselves collaborating with experts in other disciplines, such as physics, biology, and computer science. This cross-disciplinary exchange of ideas can only serve to enrich our understanding of the world around us as well as our ability to shape it for the better.

In conclusion, the application of fractal mathematics to economics offers a promising avenue for better understanding the intricate patterns that underlie economic systems. By recognizing the fractal nature of the economy and embracing the complexity it brings, we can develop more accurate models and make more informed decisions. It is our hope that the future of economics will be marked by an increasing appreciation for the beauty and power of fractals, leading to a deeper understanding of the world we live in and the way we navigate its ever-changing landscape.

Evolution and Fractals

As we embark on yet another journey into the realm of fractals, it's worth appreciating the power these fascinating structures possess in modeling complex, real-world phenomena. One might not immediately connect fractals with Charles Darwin and his groundbreaking theory of evolution, yet the undercurrents of self-similarity and scale invariance found in fractals may offer surprising insights into the biological evolution of species. In this section we traverse this unique, interdisciplinary pathway, heading toward a junction where mathematics—specifically, fractals—intertwine with biology—specifically, the theory of evolution.

The narrative of evolution, as we understand it today, speaks of a slow but steady process of change, with species gradually adapting to their environments over millions of years. This is the core principle of Darwin's theory, often visualized as a gradually sloping hill, moving ever upward toward the pinnacle of perfection.

But there is another facet of evolution that paints a more erratic, more unpredictable picture. It's a concept known as punctuated equilibrium, introduced by paleontologists Niles Eldredge and Stephen Jay Gould in the 1970s. Punctuated equilibrium proposes that species typically experience long periods of relative stability punctuated by bouts of rapid change.

This concept, though it might seem to contradict Darwin's gradualism, merely complements it. Evolution, it appears, might not be a one-size-fits-all process. And this is where fractals come into the picture. With their self-similarity and the ability to model complex phenomena, fractals offer a fresh lens through which we can interpret these patterns of life's evolution. As we shall see in the ensuing sections, fractals might just offer a robust framework to understand the punctuations and equilibriums in the grand tapestry of evolution. Strap yourselves in, folks; this is going to be one wild, fractal-fueled ride through the ebbs and flows of life's evolutionary history!

The Language of Life: DNA and Fractals: Delving deeper into the connections between fractals and evolution, we should first pay homage to the blueprint of life itself, DNA. DNA, with its elegant, double-helix structure, underpins all life forms, carrying instructions for the development, growth, reproduction, and functioning of all known organisms. If evolution is a tale, DNA is the narrator.

Yet what is less known is the fact that DNA sequences, too, exhibit fractal properties. In the early 1990s a series of studies by scientists Hamori and Ruskin found fractal patterns in noncoding sections of DNA sequences. These noncoding sections, once termed "junk DNA," turned out to be crucial in regulating gene expression. Thus, the "junk" was no junk at all!

These fractal patterns provide an extra layer of complexity to DNA and a fascinating perspective on the organization of life at a molecular level. Could it be that the fractal geometry present in our DNA reflects the complex, nonlinear processes that drive evolution? Could these patterns be the keys that unlock the mysteries of the punctuated equilibrium in evolution? The plot thickens!

And while we're here, let's not forget to mention the quirky fact that the discovery of fractal patterns in DNA sequences is not a far cry from a Dan Brown novel plot! Only, in this case, we have DNA sequences instead of ancient scriptures and fractals instead of cryptic symbols. Who knew science could be this thrilling? On to the next section we go!

Punctuated Equilibrium and Fractal Time: "Slow and steady wins the race" is a motto that does not quite fit the process of evolution. The concept of punctuated equilibrium, introduced by paleontologists Niles Eldredge and Stephen Jay Gould, proposes that evolution isn't a constant, slow process. Instead, it's marked by long periods of little change (equilibrium) punctuated by short bursts of rapid evolution. If we were to plot this on a graph over time, it would present a classic fractal pattern, echoing the ebbs and flows of evolution.

The punctuated equilibrium theory revolutionized our understanding of evolution, offering a solution to a long-standing paradox: why the fossil record often shows sudden appearances of new forms rather than gradual transformations. It turns out the answers lay in the nature of time itself.

In *Fractal Time: The Secret of 2012 and a New World Age*, author Gregg Braden suggests that time is not linear but fractal, characterized by repeating cycles and patterns. This radical theory resonates intriguingly with the concept of punctuated equilibrium, suggesting that moments of rapid evolutionary change might be part of a grand, repeating pattern on a fractal timeline.

And just when you thought it couldn't get any weirder, we add another layer of complexity: time, evolution, and fractals are wrapped together in a cosmic dance, creating a rhythm that resonates through the ages. From the DNA helix to the punctuated rhythms of evolution, the beat goes on. The symphony of life continues, and we are just beginning to discern the fractal patterns in the music. Hold on to your seats, folks.

The rollercoaster ride through the fractal landscape of evolution is far from over. Next stop—ecosystems and fractals!

We've already established the inescapable presence of fractals in the biological world, from the structure of our DNA to the pattern of evolution itself. But why stop there? Let's take a step back and look at a broader level, where life comes together in complex networks of interactions we know as ecosystems.

The balance of nature isn't merely a balance; it's a dance, choreographed to the beat of fractal rhythms. Like a spiderweb shimmering in the morning dew, the delicate intricacy of an ecosystem is mind-boggling. Species are interconnected in a network of relationships so complex that it's often impossible to predict the outcomes of even minor changes.

And yet, when we observe these systems from afar, we see familiar patterns emerging, almost like magic. The distribution of species within an ecosystem, the population dynamics, the flow of energy and nutrients—these all exhibit fractal characteristics.

Take, for example, the distribution of tree sizes in a forest. You might find many small trees, fewer medium-sized trees, and even fewer large ones. If you were to plot this distribution, it forms a curve known as a power law, which is a hallmark of fractal patterns. Or consider the branching networks of rivers and roots, the clustering of plants and animals, or the recursive loops in food webs. Once you start looking, fractals are everywhere in ecology!

In the next section, we delve into one of the most complex ecosystems known to us, the human brain, and explore the fractal patterns within. Keep your thinking cap on, folks. We're diving deep into the rabbit hole of fractal complexity!

To truly grasp the potential application of fractals to the theory of evolution, we need to first immerse ourselves in a fundamental understanding of the concept of evolution itself. At its simplest, evolution is the process through which species change over time. These changes are driven by variations that occur in the genetic material of organisms from one generation to the next, resulting in gradual but steady adaptations to changing environmental conditions.

However, not all evolutionary changes are gradual. A concept that has created fascinating debates among evolutionary biologists is the notion of punctuated equilibrium. Conceived by Niles Eldredge and Stephen Jay Gould in 1972, punctuated equilibrium proposes that evolution isn't a steady, constant creep of change. Instead, it's suggested that there are periods of relative stability punctuated by sudden bursts of rapid evolution, often triggered by significant environmental shifts or events.

In essence, punctuated equilibrium argues that species spend most of their evolutionary history in an extended state of stasis, which is then disrupted by swift periods of significant morphological change. It's a rhythm of calm punctuated by storms of genetic reshuffling and rapid adaptation. The mechanism behind these sudden spurts of evolutionary change? Mutation, genetic drift, gene flow, and natural selection acting in tandem, adapting a population to new environmental conditions with surprising speed.

Such a pattern of sudden change and stasis might seem irregular and even chaotic—but is it? Or could it reflect a deeper, underlying order? This is where the mesmerizing world of fractals could hold the key to a more nuanced understanding of evolution and punctuated equilibrium. As we continue this journey, we'll delve into how this intricate, recursive patterning may offer new insights into the very mechanisms of life's ceaseless dance of change.

Before we dive into the intriguing convergence of evolution and fractals, let's refresh our understanding of what fractals are. At their most basic, fractals are geometric figures. But these aren't just any figures—they're startlingly special, due to their defining feature of self-similarity.

Imagine a broccoli floret. Now, break off a piece. Notice anything peculiar? The smaller piece looks like a miniature replica of the larger one. Break it down again, and the pattern repeats. This is the essence of self-similarity, an intrinsic property of fractals. No matter how many times you zoom in or out, the pattern persists, a seemingly infinite cascade of repetition.

Coupled with this is the principle of scale invariance. In other words, no matter how you scale a fractal (be it by zooming in or out), it always retains its original shape. This characteristic is not only aesthetically pleasing but is also what allows fractals to model complex structures and patterns in the natural world.

From the branching of trees and the structure of our lungs to the clustering of galaxies, the world around us is saturated with fractals. They've even been used to model seemingly unpredictable phenomena, such as earthquakes and the stock market's wild gyrations, revealing surprising levels of order beneath the perceived chaos.

As we begin to consider the mechanisms of evolution and the premise of punctuated equilibrium, keep these principles of fractals in mind. The fractal lens might just bring us closer to uncovering new and profound insights into the very tapestry of life.

Since Darwin's time, the concept of evolution has commonly been imagined as a steady, gradual process—akin to the slow dripping of water that, over millennia, carves a cavernous hole in solid rock. This image of evolution as a consistent, smooth, and unhurried transformation has long held sway in our collective understanding. But is this picture entirely accurate? Not necessarily, according to the theory of punctuated equilibrium.

Picture a peaceful lake, its surface still and calm for years, decades, centuries. Then, suddenly, a stone is thrown, disrupting the tranquility with ripples and waves. Once the disturbance subsides, the lake returns to its peaceful state—but it's forever changed by the event. This, in essence, is the idea behind punctuated equilibrium.

Evidence for this theory abounds in the fossil record, which often shows species remaining virtually unchanged for millions of years, only to be replaced abruptly by markedly different species. The sudden appearance of complex life forms during the Cambrian explosion around 540 million years ago, for instance, is often cited as a case of punctuated equilibrium.

But where do fractals fit into this picture? Remember the properties of self-similarity and scale invariance. Could these fractal traits reflect the evolutionary patterns suggested by punctuated equilibrium? Could periods of evolutionary stasis and rapid change mirror the infinite cascade of repetition in fractals, writ large on the canvas of life?

As we delve deeper, we'll find that this intersection of fractals and punctuated equilibrium isn't merely a meeting of mathematics and biology, but a confluence of patterns and processes that echo across the universe, from the smallest seashell to the largest galaxy.

Imagine staring at a fractal. You notice a pattern, a particular arrangement of shapes and colors. Then you zoom in, and to your surprise, the same pattern emerges, albeit at a smaller scale. The deeper you dive, the more this pattern repeats. This property of fractals, called self-similarity, suggests that the whole can be found in its parts, again and again, ad infinitum.

Now, let's consider evolution through the lens of punctuated equilibrium. Envision a species remaining virtually unchanged for eons, its evolutionary waters calm and unrippled. Then, a sudden event—an environmental shift, perhaps, or a drastic mutation—throws a stone into these placid waters, causing a frenzy of evolutionary activity. The species transforms, adapting rapidly until it stabilizes once more, settling into another lengthy period of stasis.

At first glance, these patterns of evolution may not seem to bear the signature of fractals. But let's consider the concept of scale. Just as we can zoom into a fractal to see similar patterns at different scales, we can zoom into the evolutionary timeline. A species may experience periods of rapid evolution followed by stasis at multiple scales, from tens of years to millions.

Consider an example: a bird species might experience

a rapid evolutionary shift over a few generations due to a sudden change in climate, leading to the development of thicker plumage. Then the species remains relatively stable for several hundred years. But on a larger scale, this species could also be part of a broader evolutionary lineage that undergoes significant transformations and periods of stasis over millions of years.

In other words, the scale-invariant patterns seen in fractals can also emerge in the ebb and flow of evolutionary changes, echoing the self-similarity and scale invariance of fractals. Thus, the frenzied bursts of evolutionary adaptation followed by calm periods of stasis may not be random or erratic. Instead, they could be the expression of a deeper, fractal-like rhythm—an echo of the mathematical universe, reverberating through the living world.

Thus, the idea that fractals and punctuated equilibrium intersect becomes not only plausible but compelling, offering a fresh perspective on the ancient, ever-unfolding process of evolution.

Through this lens, each species, in its journey through time, becomes a living fractal, encapsulating the infinity of the cosmos within its DNA.

Just as our understanding of natural phenomena is shaped and clarified by creating and experimenting with models, so is our understanding of the interplay between fractals and evolution. Indeed, fractal models can bring us one step closer to visualizing the abstract concept of evolution and its intricate, multi-layered dynamics.

How might one construct a fractal model of evolution, you may wonder? One way to begin is by taking advantage of the concept of recursion, the process by which a function calls upon itself. In the world of fractals, recursion is a cornerstone, forming the basis of the self-similar patterns that unfold across scales. In an evolutionary model, recursion might represent the process of generation after generation, each calling upon the genetic "function" of the ones before.

Imagine a simple model in which an organism can

possess one of two traits: Trait A, which is beneficial in one environmental condition, and Trait B, which is beneficial in another. Each new generation could be represented as a branching point, dividing into those who inherit Trait A and those who inherit Trait B. Over time, this simple model could evolve into a fractal, illustrating the ever-branching tree of life.

Yet life and evolution are anything but simple, so our fractal models must capture this complexity. By introducing randomness, or "noise," into our models—representing factors like mutations or sudden environmental changes—we can start to simulate the unpredictability and volatility that characterizes real-world evolution.

Another layer of complexity can be added by considering not just a single trait, but a multitude of interacting traits; an interplay that creates a high-dimensional "genetic space." In this genetic space, a species' evolution might trace out a fractal pattern, crisscrossing the landscape as it adapts and adjusts to changing conditions.

In creating these models, we gain not only a visual representation of evolution but also a tool to further our understanding of how fractal dynamics might influence evolutionary processes. While still a nascent field, the exploration of fractal models in evolution promises exciting new insights into the mechanics of life's history and future.

Models, by their nature, are simplifications of reality. They cannot capture every nuance, every detail. But when they reveal patterns—patterns that resonate across species, across scales, across disciplines—then they become not just models but lenses, lenses that bring into focus the fundamental harmonies of the universe.

In the seemingly chaotic dance of evolution, we seek patterns, clues that hint at the underlying principles guiding life's endless diversification. Often, these patterns emerge in unexpected places. One such place is the field of paleontology, where ancient rocks hold the etched history of life on earth.

Paleontology, the study of fossils, provides us with

snapshots of past life forms. These remnants and impressions tell us not only about the organisms themselves, but also about the patterns of their appearances and disappearances—the rhythm of evolution. And this rhythm, much like the rhythm found in many other natural phenomena, exhibits fractal-like qualities.

But how can the fossil record, a seemingly linear accumulation of remains and impressions, be fractal in nature? To see the connection, we need to step back and look at the bigger picture. It is in this macro view that the characteristics of fractals—specifically, their self-similarity across scales—begin to manifest.

Take, for example, the sudden appearance or disappearance of a species in the fossil record: events that represent the evolutionary leaps and bounds of punctuated equilibrium. If you were to zoom in on these events, you might notice that they are not single, isolated occurrences, but rather clusters of smaller events. Much like the branches of a fractal tree, these clusters break down into subclusters, and those into sub-subclusters, exhibiting a self-similar pattern that repeats at varying scales.

The fractal nature of the fossil record extends to the temporal patterns of speciation and extinction as well. Species do not come and go at a steady, uniform rate. Rather, their appearances and disappearances ebb and flow, following a rhythm that echoes the characteristic unpredictability and scale-invariance of fractals.

Understanding these fractal patterns in the fossil record is not merely an academic exercise. It can provide crucial insights into the forces that have shaped life on earth and the dynamics that might shape its future. It may even shed light on the broad-scale structure of evolution itself, hinting at the possibility that life, in all its complexity and variety, is a grand fractal pattern etched in the fabric of space and time.

In unveiling the fractal nature of evolution, we stand on the brink of a transformative understanding of life's grand story. This profound intersection of fractal mathematics and evolutionary biology reshapes not only our comprehension of evolution's past but also its future trajectory.

Recognizing the patterns of punctuated equilibrium as fractal illuminates our understanding of evolutionary dynamics. It casts a fresh light on the vast tableau of life, revealing that the intricacies of evolution are not merely the product of random chance, but perhaps, a more complex and beautiful system unfolding according to hidden rules. This recognition of fractal patterns in evolution integrates with our knowledge of the unpredictable, nonlinear processes that shape the natural world.

From a research perspective, the incorporation of fractal models into evolutionary biology could provide a powerful new tool for scientists. By translating the convoluted processes of evolution into a fractal framework, we might develop more accurate models of how species adapt, diversify, and become extinct. For instance, understanding that species' evolution and extinction patterns follow a fractal pattern might aid in predicting future biodiversity shifts in response to climate change.

Furthermore, the fractal perspective can also help us understand the vast scale of evolutionary time. Just as fractals display self-similarity across different scales, evolutionary events might be patterned similarly whether we are looking at thousands, millions, or billions of years.

Moreover, the fractal view of punctuated equilibrium could also have profound philosophical implications. It presents a view of life as an ever-unfolding pattern, a self-replicating work of natural art. It invites us to see the biosphere not as a collection of isolated species but as a connected whole, where the microcosm reflects the macrocosm, and each part is intimately woven into the fabric of life.

In the end, the marriage of fractals and evolutionary biology heralds a potential paradigm shift, inviting us to envision evolution not just as a tree, but as an intricate fractal pattern. As we embrace this perspective, we are bound to uncover new questions, challenge old assumptions, and deepen our understanding of the captivating saga of life on Earth.

9: REAL WORLD APPLICATIONS FROM THE SMALL TO THE VERY BIG

One must still have chaos in oneself to be able to give birth to a dancing star.
—Friedrich Nietzsche

In the not-too-distant past, the twentieth century saw the birth of a new mathematical concept that would revolutionize our understanding of patterns, complexity, and chaos. This concept, known as fractals, not only traversed the boundaries of pure mathematics but also spilled over into diverse disciplines such as physics, biology, finance, and art. Today, in our advanced technological era, we find that fractals continue to make profound impacts on various fields, marking a dynamic evolution in their application and understanding since their formal introduction.

In the era of the digital revolution, the developments in fractal theory have not only been maintained but also been significantly accelerated. With the aid of modern computational capabilities, researchers now navigate the intricate universe of fractals with much more sophistication and precision than before, unveiling new facets of these mesmerizing patterns. While the fundamentals of fractals remain the same—their self-similarity, their appearance at all scales, and their intricate detail—their application has expanded dramatically in contemporary times, solidifying their role as indispensable tools in our quest to decipher the complexity of the universe.

From the quantum realms to the cosmological expanses, from the hidden layers of our DNA to the elaborate networks of social interactions, and from financial markets to the nuances of the changing climate, fractals have woven their way into the fabric of numerous scientific disciplines. The developments in the field are not just about understanding the nature of fractals better; they're about harnessing their mathematical essence to gain insights into other complex systems.

As we delve deeper into the twenty-first century, the story of fractals is far from complete. Their potential for unlocking new understandings in a variety of fields is vast. So let us explore these exciting advancements in fractal theory, revealing the intriguing, intertwined dance of order and chaos that shapes our world.

As we venture into the field of quantum physics—a realm notorious for its perplexing phenomena and mind-bending concepts—we encounter a striking synergy with fractals. The unpredictability and intricacy that define the quantum realm share a profound resemblance with the chaotic yet self-similar nature of fractals.

Quantum physics is an area where the most fundamental particles and interactions that constitute our universe are studied. Here things don't obey the familiar rules of classical physics. Instead, they follow the principles of superposition and entanglement, wave-particle duality, and a propensity for being in multiple places at once. These bizarre behaviors pose significant challenges in understanding and visualizing the quantum world.

Enter fractals. With their inherent ability to depict complexity and detail, fractals offer a unique framework for tackling the confounding complexity of quantum mechanics. Recent research has brought to light the presence of fractal structures in quantum fields and particles.

Quantum fields, for instance, can be thought of as a grid of infinitesimally small particles vibrating at different frequencies. Recent mathematical models suggest that these fields may exhibit fractal characteristics. The energy levels of these vibrating particles don't increase linearly, but rather in a way that is self-similar across different scales—a hallmark of fractals. This can help scientists better understand and predict the behavior of these fields.

Similarly, the study of quantum particles has revealed fractal-like behavior. In certain circumstances, electrons, for instance, don't move in straight or curved lines, but follow a path that resembles the intricate patterns of a fractal. This concept, known as quantum chaos or quantum fractals, is a rapidly emerging field of study, promising to shed new light on the understanding of quantum systems.

The entanglement of fractals and quantum physics paves the way for a better grasp of the complex behaviors at the universe's most fundamental level. While we are just beginning to scratch the surface, the fusion of these two seemingly disparate areas promises to lead us down a path of exciting discoveries and developments.

Fractal Cosmology

Stepping away from the subatomic world of quantum physics, we ascend to the grand scale of the cosmos. Just as fractals have found a home in the realm of the very small, they also appear to have a profound connection to the very large—the universe itself. This brings us to the fascinating concept of fractal cosmology.

In the twentieth century, our understanding of the universe was largely based on the cosmological principle. This principle assumes that, on a large enough scale, the universe is homogeneous and isotropic—meaning it looks the same everywhere and in every direction. But the advent of powerful telescopes and advancements in cosmological studies have given rise to observations that seem to challenge this principle.

As astronomers charted the positions of galaxies and galaxy clusters, they noticed something intriguing: these celestial bodies don't appear to be randomly sprinkled throughout the cosmos. Instead, they seem to follow a complex, irregular pattern that's highly reminiscent of a fractal.

In this pattern, clusters of galaxies form larger superclusters, which in turn form even larger structures, each repeating the clustering pattern at a larger scale. This self-similar pattern across different scales, from individual galaxies to the vast cosmic web, is strikingly similar to the structure of fractals.

This concept of a fractal universe holds profound implications for cosmological theory. It questions our assumptions about the large-scale structure of the universe, shaking the foundations of the standard cosmological model. It also offers a potential explanation for some of the mysteries of cosmology, such as the nature of dark matter and dark energy, which comprise a large portion of the universe's mass and energy but remain poorly understood.

However, fractal cosmology is still a young and controversial field. Many questions remain, and more data is needed to confirm or refute the fractal nature of the universe. Regardless of the outcome, the exploration of fractal cosmology is pushing the boundaries of our understanding and challenging us to see our universe in a new light—through the intricate, self-similar lens of fractals.

Fractal cosmology is a relatively recent field of study that examines the possibility that the large-scale structure of the universe follows a fractal pattern. This is a significant departure from traditional cosmological thinking, which often assumes that the universe is uniform and homogeneous on the largest scales.

The concept of fractals, as developed by mathematicians such as Benoît Mandelbrot, has led to groundbreaking discoveries in various fields of science. In fractal cosmology, researchers apply these principles to the study of celestial bodies and their distribution in the universe. This emerging field has challenged established ideas about the nature of space, time, and the fundamental forces governing the cosmos.

Fractal cosmology is grounded in the observation that the distribution of galaxies and galaxy clusters appears to exhibit self-similar patterns. These patterns, reminiscent of fractals, suggest that the universe may have a noninteger, fractal dimension on large scales. This idea has prompted researchers to reconsider the assumptions and models underlying our understanding of the cosmos.

In the following sections, we will delve into the concept of fractal cosmology, exploring its origins, implications, and controversies. We will examine the current state of research in this area, consider its connections to other fields such as quantum physics, and discuss the future of fractal cosmology as scientists continue to probe the depths of the universe.

The Idea of a Fractal Universe: The idea of a fractal universe grew out of observations of the large-scale structure of the universe. When astronomers look at the distribution of galaxies and galaxy clusters in the universe, they don't see a smooth, uniform spread. Instead, galaxies tend to cluster together, forming intricate structures such as filaments and walls that encircle vast, empty voids. This arrangement appears to exhibit self-similar patterns, a key characteristic of fractals.

Applying this concept to cosmology, some researchers hypothesize that the universe's large-scale structure may also be fractal in nature. This means that whether we are looking at a region of space spanning hundreds of millions of light-years or just a few million, we should see similar patterns in the distribution of matter.

However, the idea of a fractal universe is not without controversy. Traditional cosmology holds that on the largest scales, the universe is homogeneous and isotropic—the same in all places and all directions. This principle, known as the cosmological principle, is a cornerstone of our current understanding of the universe. The fractal universe concept seems to contradict this principle, suggesting instead that the universe is not the same on all scales.

The question of whether the universe is truly fractal remains open. It is a complex issue that touches on deep, fundamental questions about the nature of the universe and our understanding of cosmology. The search for answers continues to drive research in this fascinating and provocative field.

Fractal Geometry and the Universe: In the previous section, we explored the provocative idea of a fractal universe—a cosmos where the distribution of matter mirrors the self-similar patterns characteristic of fractals. In this section, we will delve further into the connection between fractal geometry and cosmology.

The heart of fractal geometry lies in the concept of self-similarity. This is the idea that an object is made up of smaller copies of itself, each scaled by a certain factor. This leads to intricate structures that look similar at any level of zoom. The Mandelbrot set, a fractal that has captivated the imaginations of mathematicians and laypeople alike, is a prime example of such a structure.

Now if we apply this idea to the universe, we end up with a model of cosmos where clusters of galaxies are merely "cosmic coastlines" with an intricate, possibly infinite, structure that repeats itself at different scales.

Yet the universe doesn't just consist of matter. There's also dark matter, an invisible substance that interacts with the rest of the universe through gravity. Observations suggest that dark matter forms a cosmic web with galaxies forming at the nodes where the web's filaments intersect. This cosmic web also appears to exhibit fractal-like patterns.

Fractal geometry also offers a new perspective on the evolution of the universe. In the standard model of cosmology, the universe started with a Big Bang and has been expanding ever since. Yet if the universe is indeed fractal, this might imply a different kind of cosmic evolution, one where new structures constantly emerge on smaller and smaller scales.

The idea of a fractal universe is both mesmerizing and challenging. It adds another layer of complexity to our understanding of the cosmos, inviting us to reconsider our notions about space, time, and the very nature of reality.

Fractals and the Cosmic Web: Having established some preliminary groundwork, it's time to delve deeper into one of the most intriguing aspects of a potential fractal universe: the cosmic web. This interconnected system of galaxies, intergalactic gas, and dark matter weaves an intricate tapestry across the vast expanse of the cosmos, bearing a striking resemblance to the endlessly complex patterns found within fractals.

The cosmic web theory emerged as a direct consequence of our current understanding of dark matter. This unseen substance constitutes around 85 percent of the universe's mass, and its gravitational pull is the primary sculptor of cosmic structures. Galaxies are not randomly scattered across the universe; instead, they congregate along dense regions of dark matter, interconnected by vast cosmic filaments. Where these filaments intersect, we find vast clusters of galaxies.

When we map out the distribution of galaxies, we find a structure that is decidedly nonrandom. Galaxies cluster into filamentlike structures with large voids in between. This distribution has a striking resemblance to fractal patterns. Like the branching patterns seen in trees or veins, galaxies distribute in a manner that suggests self-similarity over multiple scales.

Many researchers have explored this fractal nature of cosmic structures. They have created three-dimensional maps of the universe that illuminate this vast cosmic web in unprecedented detail, highlighting the filaments and voids that constitute our universe's large-scale structure. This work has provided strong evidence for the fractal nature of the universe on certain scales, although there's much debate about whether this fractality extends to all scales.

Such a vision of a fractal cosmic web is both fascinating and challenging. It reshapes our understanding of the universe, suggesting that its structure may be governed by a beautiful, complex geometry that mirrors the intricacies found within fractals. It's a testament to the intertwined nature of the cosmos, a literal and metaphorical web that interlinks every celestial body within it. This realization underscores our deep connection to the cosmos, reminding us that we're an integral part of this grand cosmic web.

Fractals and Inflationary Cosmology: The inflationary model is a cornerstone of modern cosmology. It proposes a period of extremely rapid expansion of the universe, known as cosmic inflation, in the fractions of a second following the Big Bang. This idea has revolutionized our understanding of the universe's origins and evolution, helping to explain the uniformity of the cosmic microwave background radiation, the distribution of large-scale structures, and the flatness of the universe.

In recent years, intriguing connections between fractals and inflationary cosmology have been unearthed. In particular, fractal mathematics have been used to model the quantum fluctuations that were stretched to cosmic scales during inflation, giving rise to the seeds of structure in the universe.

According to the inflationary model, the universe at its earliest moments was subject to quantum fluctuations. These fluctuations, while initially small and local, were dramatically stretched during the inflationary period to macroscopic scales, seeding the formation of galaxies and clusters of galaxies. The spatial distribution of these seeds, and, by extension, the cosmic structures they gave rise to, display a degree of self-similarity and scale-invariance—properties reminiscent of fractals.

Furthermore, inflationary cosmology has opened the door to the concept of a multiverse—a multitude of universes where our observable universe is just one among many. Some models propose a "fractal multiverse," where new universes are continually spawned within black holes of parent universes. This process creates a sort of self-similar, infinite tree of universes, where each branching point signifies the birth of a new universe.

Such concepts are highly speculative and push the boundaries of our current scientific understanding. However, they serve as a testament to the potential reach of fractal mathematics in shaping our understanding of the cosmos. It's a testament to the versatility and power of the fractal concept that it finds applicability in these explorations of the very edge of human knowledge.

Fractal Structures in Cosmic Microwave Background Radiation: The cosmic microwave background (CMB) radiation is a snapshot of the universe when it was just about 380,000 years old. It's a relic from the Big Bang, offering a wealth of insights into the universe's birth and evolution.

The CMB is remarkably uniform, with tiny temperature fluctuations that correspond to the density variations in the early universe. These fluctuations are the seeds from which galaxies and clusters of galaxies eventually formed. As such, the study of these tiny variations is fundamental for understanding the large-scale structure of the universe.

Recent research has suggested that these small variations in the CMB might exhibit a degree of fractal behavior. The patterns of the temperature fluctuations in the CMB, while not strictly self-similar, do show some scale-invariant characteristics. This could imply that the processes that generated these fluctuations—which took place during the inflationary period of the universe—were fundamentally fractal in nature.

The research into fractal patterns in the CMB is still in its early stages, and there is much more to be explored. But the initial findings offer tantalizing hints of yet another area where the reach of fractals might extend, revealing more about the universe's formation and structure.

One of the future tasks for the next generation of CMB experiments is to verify whether these patterns in the temperature fluctuations do indeed follow a fractal distribution. If confirmed, it would offer a new tool for understanding the universe's early moments and the processes that led to the formation of large-scale cosmic structures.

The Future of Fractal Cosmology: Promises and Challenges: Fractal cosmology is undoubtedly a burgeoning field of study. It is offering a new lens to view the universe, one that is shedding light on its most intricate and large-scale structures. As we continue to develop and refine our understanding of fractals and apply them to cosmology, we are certain to gain novel insights into the workings of the universe.

One promising area of future study is in the realm of dark matter and dark energy. These two mysterious components make up approximately 95 percent of the universe, and their nature is one of the most significant unsolved problems in physics. It has been suggested that the fractal distribution of matter could offer insights into these enigmatic phenomena. If the large-scale structure of the universe indeed follows a fractal pattern, this could have profound implications for our understanding of dark matter and dark energy.

Moreover, the advent of even more powerful telescopes and observation tools will enhance our capacity to study the universe's large-scale structure. Upcoming missions like the James Webb Space Telescope and Euclid, a European Space Agency mission, will deliver a wealth of new data, offering unprecedented opportunities to test and refine the theories of fractal cosmology.

However, fractal cosmology is not without its challenges. One of the significant criticisms is the matter of defining and measuring scale. Given that fractals are fundamentally about self-similarity across scales, understanding what constitutes a scale in cosmology and how to measure it accurately is critical. Additionally, the introduction of fractals into cosmology necessitates a rethinking and retooling of standard cosmological models, which is no small feat.

In the end, the future of fractal cosmology is ripe with promise. It's a discipline that's poised to not only deepen our understanding of the cosmos, but also continue to reaffirm the ubiquity of fractals, from the tiniest particles to the vastness of the universe. Whether fractal cosmology will ultimately revolutionize our understanding of the universe is yet to be seen, but it surely holds the potential to leave a lasting impact on the field.

Conclusions and the Future of Fractal Cosmology: As we've seen, the application of fractal theory to cosmology offers a truly transformative lens through which to view our universe. From the potential of the holographic principle to the understanding of dark matter distribution, the self-similarity principles of fractals bring new light to some of the most puzzling questions in our cosmos.

Yet the world of fractal cosmology is still very much a field in flux. Current research is expanding and challenging our previous knowledge, reshaping our cosmic perspective. The James Webb Space Telescope, with its advanced capabilities, offers the potential to further our understanding and confirm or deny the many theories in fractal cosmology.

Looking forward, as we continue our exploration of the universe, we can anticipate that fractal models will evolve and become more refined. With these advancements, our comprehension of the universe will undoubtedly deepen. Fractal cosmology isn't just about understanding the structure of the cosmos; it's about revealing the fundamental principles that guide its formation and evolution.

In conclusion, the universe's fractal nature is a testament to the remarkable interplay of simplicity and complexity, chaos and order that characterizes our reality. As we continue to probe the cosmos and uncover its secrets, the fractal perspective will continue to be an invaluable tool, forever changing how we perceive our place within this grand cosmic design.

Fractal cosmology, then, isn't just a field of study—it's a new way of viewing and understanding the cosmos. As we look to the future, one thing is clear: the universe has a lot more fractal secrets to reveal. And as we unlock these secrets, we will undoubtedly gain a deeper understanding of our place in this intricate, beautiful cosmic pattern.

Fractals in Medicine: A New Frontier in Medical Science
If you were to ask someone what mathematics has to do with medicine, you might receive a puzzled look. Yet as we've learned throughout this book, mathematics—and particularly fractal mathematics—offers unique insights into the natural world. The world of medicine is no exception.

Fractals, with their unique ability to describe complex structures and patterns, have begun to play a transformative role in various fields of medical science. They offer a new language to interpret the intricate patterns found in the human body, whether it's the branching of our lungs or the web of neurons in our brains.

For instance, medical imaging is one area where fractals have shown great promise. Fractal analysis is increasingly being applied to improve the interpretation of images from MRI scans, CT scans, and other imaging technologies. This helps to identify abnormal structures or patterns that may signal disease. For example, tumor growth often follows a fractal pattern, making fractal analysis a potentially powerful tool in the early detection of cancers.

Diagnostics, too, have seen the incorporation of fractals. They have been used in the analysis of heart rate variability, a measurement that has proven valuable for predicting cardiac events. The seemingly random fluctuation of heartbeats is, in fact, a complex, fractal rhythm. Traditional linear methods of analysis can overlook this complexity, but fractal analysis can uncover subtle changes that may indicate a problem.

Moreover, understanding physiological functions and their fractal nature has given us a new perspective on health and disease. It's becoming apparent that many biological systems, from our circulatory system to our nervous system, exhibit fractal properties. When these systems deviate from their fractal nature, it could be a sign of disease or dysfunction.

Fractals, therefore, are not just mathematical curiosities. They're powerful tools that are helping to transform medicine, enhancing our understanding of human physiology, improving diagnostics, and potentially even guiding treatment. As we move forward into the future of medicine, the role of fractals is set to become even more significant.

Chaos Theory and Fractals: Unraveling Complexity

Chaos theory has seen significant advancements in recent years, and these developments have led to a deeper understanding of fractal patterns. A chaotic system, by definition, has a deterministic yet unpredictable behavior. It can produce a seemingly random and disorderly output from a set of deterministic laws. Yet when we map out these outputs, an intricate, never-repeating, but infinitely self-similar pattern emerges—a fractal!

For instance, consider the logistic map, a simple mathematical model used to describe population dynamics in ecology. The map takes a population level at a given time and uses it to predict the population level in the next period. When certain values are used in this model, the population levels fluctuate chaotically. If we plot the population levels over time, we see a bifurcation diagram that is an exquisite fractal.

In other words, the fingerprint of chaos is a fractal. They are not just random or disordered systems, but a complex interplay of order and disorder, captured best by the infinite complexity of fractals. Scientists are just beginning to tap into the potential of using fractals to better understand and predict chaotic systems. From weather systems to stock markets, the applications are as boundless as the fractals themselves.

Fractal Antennas: Revolutionizing Wireless Communication

In the modern age of smart devices and wireless communication, fractals have found a unique application in the design of antennas. You may wonder, how can a mathematical concept related to snowflakes, coastlines, and fern leaves help improve your Wi-Fi signal? The answer lies in the fascinating properties of fractals.

Fractal antennas, as the name suggests, are antennas designed using fractal shapes. These intricate, self-similar shapes allow for antennas to be smaller, yet still powerful. Fractals have this peculiar property of "space-filling." They can increase the length or area within a given space without increasing the space itself. Applying this to antenna design means we can have antennas that are physically small but have a much larger effective length.

Moreover, the self-similarity property of fractals allows these antennas to operate at different frequencies simultaneously. A single fractal antenna can receive a broad range of signals, eliminating the need for multiple antennas tuned to different frequencies.

This technology is still evolving, but it already has exciting applications. Fractal antennas are now being used in cell phones, Wi-Fi routers, and other wireless devices. They could even pave the way for more efficient, more reliable wireless communication networks in the future. These tiny, intricate antennas represent another way that the infinite complexity of fractals can be harnessed for practical, everyday use.

Fractals and Machine Learning: AI Meets Infinity

In the ever-evolving world of artificial intelligence (AI) and machine learning, it's not uncommon for researchers to borrow ideas from different fields. Lately, they've been delving into the intricate, infinite world of fractals to make breakthroughs.

Fractals, with their self-similarity and intricate structures, have a surprising parallel in the realm of machine learning. Take neural networks, for instance. These are computer systems loosely modeled on the human brain, designed to recognize patterns. They learn from vast amounts of data by adjusting the strength of connections between artificial neurons. Just as fractals are composed of small, self-similar units that build up to a complex whole, neural networks consist of layers of interconnected nodes, each performing a simple computation that contributes to the network's overall output.

Interestingly, fractals can help optimize these neural networks. Researchers have started using fractal structures to design more efficient and powerful neural networks. These fractal neural networks exploit the self-similarity of fractals to create networks that can scale up in complexity without a similar increase in computational costs.

Furthermore, machine learning algorithms can also use fractal analysis to identify patterns and make predictions. For example, fractal dimensions, a measure of complexity in a data set, can be used to analyze and classify complex data, such as satellite imagery or medical scans. Fractal-based machine learning models have been used to detect diseases from medical images, predict stock market trends, and even analyze geospatial data for urban planning.

But we're just scratching the surface. The fusion of fractal theory with machine learning is a promising frontier that could revolutionize how AI learns and evolves. After all, when infinite complexity meets artificial intelligence, the possibilities are, quite fittingly, endless.

Fractal Antennas: The Shape of Communication

In the world of telecommunications, we've seen a shift from large, cumbersome devices to smaller, more efficient ones. The drive toward miniaturization, however, isn't just about the size of the devices we hold in our hands, but also about the components that allow these devices to communicate. Enter fractal antennas.

Simply put, antennas are devices that convert electrical signals into radio waves and vice versa. Traditional antennas are often large, and their size is directly related to the wavelength of the signal they are designed to handle. This is where fractals, with their magic of self-similarity and space-filling properties, come in.

Fractal antennas, as the name suggests, are antennas that employ fractal shapes in their design. The key advantage of using fractals in antenna design is their ability to receive and transmit at multiple frequencies simultaneously. The self-similar nature of fractals means that a fractal antenna can be designed to operate at multiple wavelengths, making them multifrequency or wideband antennas. In other words, one fractal antenna can do the work of several traditional antennas.

In addition to their wideband capabilities, fractal antennas also have other attractive qualities such as being more compact and efficient. Thanks to their fractal designs, these antennas can be made smaller without sacrificing performance. They also demonstrate enhanced efficiency and better signal reception, partly due to the infinite perimeter of fractals that help in capturing more signal.

In a world that increasingly depends on wireless communication, from smartphones to satellites, these advantages make fractal antennas a critical development. They're used in a variety of applications, including in cellular networks, in GPS systems, and even in space exploration. As we move forward into a world of 5G, IoT, and beyond, fractal antennas may just be the perfect emblem of modern telecommunications: small in size, vast in capability, and infinitely fascinating.

Climate Modeling and Fractals: Predicting the Unpredictable
As we grapple with the challenges of climate change, accurate climate modeling becomes more critical than ever. Yet traditional models often fall short of capturing the inherent complexity and scale-dependent nature of weather systems. That's where fractals come in.

Weather and climate systems exhibit an inherent self-similarity, reminiscent of fractal structures. Whether we look at cloud formations, the branching of lightning, or the pattern of winds and ocean currents, we see patterns repeating at multiple scales. Recognizing this, researchers have turned to fractal mathematics to improve the accuracy and predictive power of climate models.

In the context of climate modeling, fractals help in several ways. First, they allow us to represent natural phenomena that have noninteger dimensions, such as coastlines, in a more accurate manner. This provides a more realistic representation of these features within climate models, which in turn improves the models' performance.

Second, fractals help in capturing the inherent unpredictability of weather systems. Traditional models often struggle with capturing sudden shifts in weather patterns, a characteristic feature of chaotic systems. But with fractals, researchers can model these shifts more accurately, thanks to the intricate patterns fractals can generate even with simple mathematical rules.

Third, fractals can help in downsizing computational load. By using fractals, researchers can generate detailed, realistic representations of complex systems with simpler mathematical operations, making the process more efficient.

In essence, fractals offer a new lens through which to view, understand, and predict our ever-changing climate. By leveraging the power of fractals, scientists are unraveling the complex patterns of our planet's climate system, enhancing our ability to predict future climate scenarios and empowering us to take more effective actions to mitigate the impacts of climate change. The climate challenge is a grand one, but as we're learning, sometimes the answer lies in the smallest patterns.

Fractals in Modern Art and Architecture: Breaking Boundaries
Art and architecture have always been inspired by the beauty and complexity of the natural world. It's no surprise, then, that the discovery of fractals—mathematical structures that mimic the self-similar patterns seen in nature—has made its way into these fields.

In the art world, the fascinating visuals created by fractals have been embraced enthusiastically. Digital artists manipulate mathematical formulas to generate incredibly intricate and beautiful fractal images, marrying artistry with mathematics in a union that blurs the line between science and art. The resulting artworks captivate viewers with their infinite complexity and ethereal beauty, providing a visual journey into the abstract realms of mathematical landscapes.

In architecture, the use of fractals has revolutionized design approaches. Traditional design principles often emphasized balance and symmetry, but the incorporation of fractal principles has introduced a new dimension of complexity and beauty. Fractal architecture, which mimics the irregular yet self-similar patterns seen in nature, results in buildings that not only are aesthetically appealing but also blend seamlessly with their natural surroundings.

Take, for instance, the design of certain modern buildings that mimic the fractal branching patterns of trees, enabling the structure to balance weight more efficiently while also offering a pleasing aesthetic that mirrors the natural world. Some urban planners even suggest that city layouts designed with fractal principles might improve living conditions by mimicking the scale-variance seen in nature.

Fractals have thus reshaped our understanding of aesthetics, inspiring us to push the boundaries of art and design. By incorporating the principles of self-similarity and infinite complexity, modern artists and architects have embraced the fractal revolution, demonstrating that science and art, far from being separate domains, can intermingle to create something truly magnificent.

Fractal Generation and Computer Algorithms: Pushing Computational Boundaries

As we've seen, fractals have applications spanning various fields, but it's within the realm of computer science that they arguably shine the brightest. Fractals, with their inherent recursive nature and infinite complexity, provide fertile ground for testing and advancing computer algorithms.

At the core of fractal generation is the concept of recursion, a principle that underpins many computer algorithms. Recursive algorithms are those that solve a problem by solving smaller instances of the same problem. They are particularly useful when dealing with fractals, as fractals are essentially an infinitely repeating pattern that occurs at every scale.

Fractal algorithms—often written in programming languages like Python or Java—begin with a basic shape and then modify it in a specific way, over and over again. These algorithms are powerful tools for both creating fractal graphics and exploring their properties.

Advancements in computational power and graphics processing have also made real-time rendering of complex fractal landscapes possible, pushing the boundaries of what can be visualized and explored. Researchers are now able to delve deeper into the Mandelbrot set and other fractals, unveiling previously unseen territories in these mathematical landscapes.

Moreover, fractals and their algorithms have found practical applications in data compression. Since fractals are mathematically efficient representations of data, they can be used to compress image and sound files to a fraction of their original size without significant loss of quality.

Finally, fractals and computer algorithms have facilitated advancements in chaos theory, a branch of mathematics that studies complex systems whose behavior is extremely sensitive to slight changes in conditions. Fractals offer a way to visually represent chaotic systems, aiding in the study and understanding of their unpredictable yet deterministic nature.

So while fractals may be abstract mathematical constructs, their real-world applications, powered by advancements in computer science, are very much tangible and impactful. They truly exemplify the interplay between mathematics, science, technology, and art.

Fractals in Environmental Modeling: Embracing Complexity
From the microscopic level to the macroscopic, the natural world is filled with complexities that challenge the limits

of human understanding. Enter fractals, the mathematical structures renowned for their ability to capture complexity, and it becomes apparent how they can transform our approach to environmental modeling.

At the heart of environmental modeling is the task of representing the intricate and complex processes of nature within a mathematical framework. These processes range from the growth of coral reefs to the branching of rivers to the spread of forest fires. Traditionally, these were approximated using simple geometrical shapes or regular patterns, but such models often fell short in capturing the inherent complexity of natural phenomena.

Fractals, however, with their self-similar patterns and intricate detail at all scales, offer a powerful alternative. They provide a language to describe the irregular shapes and patterns observed in nature more accurately.

For example, consider the fractal nature of river networks. The branching pattern of a river system is a self-similar structure that can be modeled using fractals. Understanding this fractal structure can provide valuable into predicting water flow, sediment transport, and the impact of environmental changes.

Similarly, the growth patterns of forests, particularly the spread of forest fires, have been modeled using fractal geometry. A forest fire, once ignited, doesn't spread uniformly. It advances in a patchy, irregular manner, dictated by factors such as wind, humidity, and the spatial distribution of vegetation—a pattern well-suited to fractal modeling.

The application of fractals in environmental modeling is a testament to their versatility and efficacy in representing complex systems. With each passing year, as more data is collected and computational power increases, we can expect to see an even greater role for fractals in helping us to understand, predict, and respond to the complex environmental challenges that lie ahead.

Fractals and Blockchain: From Cryptography to Cryptocurrency
With the advent of blockchain technology and cryptocurrencies like Bitcoin, fractal geometry has found a new, surprising application area. Blockchain, the underlying technology behind cryptocurrencies, is a decentralized ledger of all transactions across a peer-to-peer network. This technology is not only transforming the world of finance but also offering novel ways to look at and implement fractal patterns.

The blockchain itself is an inherently fractal structure. A blockchain consists of multiple blocks of information, each linked to the one before it, creating a chain. Like a fractal, each block is a self-contained unit of information that is part of a larger, similar whole. The block contains a history of itself and all the preceding blocks, similar to how a fractal pattern contains a scaled-down version of the whole.

Moreover, the hashing functions used in blockchain, which turn an input of any length into a fixed output, bear a striking resemblance to the iterative processes used to generate fractals. These functions take an input and transform it into an output, and then use that output as the input for the next iteration, much like how a fractal equation takes a number, performs operations on it, and then feeds the result back into the equation.

In the context of cryptocurrencies, the value fluctuations of digital assets like Bitcoin also exhibit fractal patterns. Market trends in cryptocurrency often mirror themselves at different scales, similar to the scale-invariance of fractals. Traders and analysts use this property to make predictions and identify market trends.

The intersection of fractals and blockchain represents an exciting frontier in the world of mathematics and technology. As our understanding of fractals deepens and technology continues to evolve, it is likely that more applications of fractals in blockchain and cryptocurrency will emerge. The self-replicating patterns of fractals may hold the key to new breakthroughs in cryptography, network design, and decentralized systems.

Conclusion: The Future is Fractal

From the subatomic particles to the vast cosmos, from the branching veins of leaves to the robust networks of the internet, the theme of fractals has proven to be universal. This fascinating concept has transcended its mathematical roots, permeating various fields from quantum physics to geology, from medicine to artificial intelligence. Today, as we stand on the precipice of new discoveries and inventions, fractals continue to hold a prominent place, shaping the way we understand and interact with our world.

Our exploration in this chapter has taken us on a journey across diverse landscapes of knowledge. We've seen how quantum fields hint at a fractal structure, opening new possibilities in the realm of the smallest particles. We delved into the fractal cosmos, reflecting on the patterns that might guide the structure of our universe. We looked at how medicine uses fractal analysis to diagnose and understand diseases, and how cutting-edge technology, like AI and fractal antennas, is making use of this concept to revolutionize their domains.

Yet this is not the end. Instead, we find ourselves at the beginning of a new chapter. As technology continues to evolve and our knowledge expands, we are sure to find more and more applications of fractals. They will continue to inspire, challenge, and guide us into the future. As we move forward, it's clear that the world of fractals is far from being fully explored, promising rich rewards for those daring enough to delve into its depths.

As we close this chapter, let's remember a crucial aspect: the beauty of fractals is not confined to their applications or their mathematical definitions. The true allure lies in their inherent ability to capture complexity and chaos, to hold infinity within a finite space. They remind us of the intricate, interconnected nature of our universe, challenging us to look closer, to question more, and to marvel at the stunning patterns dancing around us. After all, we live in a fractal world, and the future, undoubtedly, is fractal.

10: CONCLUSION

Mathematics is the queen of sciences and number theory is the queen of mathematics. She often condescends to render service to astronomy and other natural sciences, but in all relations she is entitled to the first rank.
—Carl Friedrich Gauss

The Allure of Fractals

Fractals are mathematical chameleons, ever changing and endlessly complex. They weave their intricate patterns across the canvases of science, art, nature, and human creations, quietly embedding their exquisite forms into the fabric of our lives. The allure of fractals, it seems, lies not just in their objective beauty, but in the incredible depth of their simplicity.

Our journey through the fascinating world of fractals began with the basics: the definition of a fractal and its foundational mathematical principles. We learned how these infinitely complex shapes are born from a deceptively simple process of iteration, replicating themselves at every scale. Just like a hall of mirrors that reflects an image ad infinitum, fractals encapsulate the concept of infinity within a finite form. This, in itself, is a profoundly beautiful idea.

Fractals are also innately universal, crossing disciplinary boundaries with ease. Whether we're discussing the patterns found in a leaf's veins or the distribution of galaxies across the cosmos, fractals are present. This universality holds a certain charm; it offers a visual representation of the interconnectedness of all things.

Their universality, however, is not their only appeal. Fractals are more than just patterns. They represent a fundamental shift in how we perceive and interact with the world around us. They challenge our traditional understandings of geometry and force us to reconsider our perceptions of shape, dimension, and scale.

Throughout this journey, we've seen how fractals have inspired and influenced a variety of fields, from visual arts and music to computer algorithms, physics, and biology. Their captivating patterns and infinite complexity offer an endless source of inspiration, a testament to the beauty inherent in the language of mathematics.

As we conclude our exploration, we recognize that the allure of fractals extends far beyond their aesthetic appeal. Fractals remind us of the astonishing beauty that can emerge from simple rules and processes. They highlight the underlying patterns that stitch together the fabric of the universe. Above all, they represent a celebration of the complexity, elegance, and, indeed, the inherent magic of the mathematical world.

The Influence of Fractals

Fractals have not just inspired us, they have also shaped and influenced a wide range of disciplines, setting new paths for exploration and study. This influence is as varied and expansive as the fractals themselves.

In the realm of mathematics, fractals have challenged us to rethink our definitions of dimensions and shapes, presenting a form that blends both dimensions and that evolves with self-replicating patterns. The complex geometry of fractals has provided new mathematical tools and perspectives, expanding our understanding of the mathematical landscape.

The influence of fractals extends far beyond mathematics. In physics, the concept of fractal dimensions is used to describe the statistical distribution of galaxies in the universe. Quantum mechanics has found fractals in the probability clouds of particle distributions. These discoveries have deepened our understanding of the cosmos and the fundamental particles that compose it.

Biology has been similarly touched by fractals, with their patterns observed in the growth and branching of trees, the structure of blood vessels, and even the rhythm of our hearts. Understanding these fractal patterns can provide insights into biological growth, structure, and function.

Perhaps one of the most unexpected spheres of fractal influence has been in the world of art and design. Artists have embraced the beauty and complexity of fractals, using them to create visually stunning works. The same fractal algorithms that help us understand galaxy distributions or tree growth are now used to create breathtaking computer graphics and animations.

Moreover, the fields of computer science and data analysis have been revolutionized by fractal algorithms. Fractal compression algorithms allow us to store large amounts of information in a compact form, making the digital world more efficient. Machine learning and artificial intelligence have also benefited from fractal theory, with fractal-based techniques used in pattern recognition and data classification.

In sum, the influence of fractals is vast and far-reaching. They have challenged us, fascinated us, and ultimately enriched our understanding of the world around us. The story of fractals is a testament to the power of mathematical ideas to inspire and transform, to influence and create, and to reveal the beauty hidden within the complexity of our universe.

The Future of Fractals

As we gaze into the fractal future, it becomes increasingly clear that this intriguing mathematical concept will continue to weave its intricate web into various scientific disciplines, as well as into our everyday lives. The future of fractals is dynamic, promising, and, fittingly, infinitely complex.

At the forefront of scientific research, fractals are contributing to our understanding of complex systems. In the realm of quantum physics, researchers are investigating how fractal geometry could potentially reshape our understanding of quantum field theory and the nature of the universe. This not only could revolutionize quantum mechanics, but it could also affect our fundamental understanding of reality.

In medicine and biology, fractal analysis continues to provide valuable insights. From diagnosing diseases through the fractal analysis of medical images to understanding the fractal nature of our DNA and cell structures, the applications are immense. Fractals could also shed light on the complexities of brain function, contributing to advancements in neuroscience and mental health.

On a larger scale, fractal algorithms are becoming more sophisticated and are being applied to simulate complex natural phenomena, such as the formation of galaxies or the evolution of weather patterns. These algorithms could enable us to create more accurate models, improve predictions, and develop better strategies for dealing with global challenges, such as climate change.

In the realm of technology, fractals are being used to advance fields like artificial intelligence and machine learning. Fractal-based techniques are aiding pattern recognition, data analysis, and the development of neural networks that mimic the complexity of the human brain.

In the art and design world, the beauty and complexity of fractals will continue to captivate and inspire artists, leading to innovative creations that blend science and art in stunning ways. From digital art to architecture, the aesthetic appeal and unique properties of fractals provide an endless source of creativity.

The future of fractals holds infinite possibilities, much like the fractals themselves. As we delve deeper into this fascinating realm, we can expect to uncover more of its secrets and continue to be awed by its intricate beauty. The journey into the world of fractals is far from over; in fact, it's only just begun.

Reflections on Fractals and Complexity
In the journey through the mesmerizing world of fractals,

one of the most striking realizations is the complexity that underlies even the simplest forms in nature and mathematics. This section provides an opportunity for introspection, an exploration of the philosophical implications that arise from our understanding of fractals and the inherent complexity of our universe.

A fundamental understanding of fractals encourages a shift in perspective. The concept of self-similarity across scales prompts us to consider how patterns at one level of complexity can inform our understanding of patterns at another. This perspective is not just limited to mathematics or physical systems; it extends to how we view societal structures, economic systems, or even the nature of consciousness itself.

Fractals remind us that complexity can arise from simple rules. A simple iterative process can give rise to infinite complexity and unpredictable outcomes. This is evident not only in the generation of intricate fractal patterns but also in real-world phenomena such as the weather system or stock market fluctuations.

This understanding of complexity has profound implications. For instance, in our increasingly interconnected world, it helps us understand the potential cascading effects of small changes—a concept crucial to fields ranging from ecology to economics to global politics.

Moreover, our exploration of fractals invites us to ponder over the beauty of mathematics and its uncanny ability to describe the natural world. The ubiquitous presence of fractal patterns in nature is a testament to the profound connection between mathematical structures and the physical universe.

Finally, understanding fractals is a humbling reminder of our limited human perspective. Fractals, with their infinite complexity and detail, symbolize the vastness of the universe and the endless depth of knowledge waiting to be discovered. As we stand on the brink of this mathematical abyss, we are prompted to recognize both the vastness of our ignorance and the limitless potential for learning and growth.

In essence, our journey through fractals has been not just an exploration of a mathematical concept but a profound meditation on complexity, pattern, beauty, and the interconnectedness of all things.

The Future of Fractals: A Peek into the Uncharted Territory
In the last section of this grand exploration of fractals, we venture into the realm of what lies ahead. The world of fractals, despite its rich history and significant advancements, remains a field ripe with exciting and unexplored possibilities.

So, where do we go from here? The short answer is anywhere and everywhere. Fractals have demonstrated their profound versatility, appearing in areas we never thought possible. As we advance in technology and science, new avenues for their application will undoubtedly emerge.

One exciting frontier is in the field of quantum computing. How might the properties of fractals inform the development of quantum algorithms or the design of quantum systems? There's fertile ground here for exploration.

Similarly, the field of artificial intelligence, particularly neural networks, could also benefit from a deeper integration of fractal theory. Given that fractal patterns appear to model learning processes and the development of complexity in natural systems, they may provide insights into how to construct more efficient and powerful learning algorithms.

The exploration of space may also benefit from fractals. As we probe the cosmos with increasingly sophisticated technology, we may discover that fractal structures are not confined to our earthly environment but are a universal phenomenon.

Furthermore, there is a deep and abiding beauty in fractals that resonates with us on a human level. This aesthetic quality, combined with their mathematical properties, make them a source of endless inspiration for artists, musicians, architects, and designers.

In essence, fractals will continue to serve as a

bridge linking diverse fields, inspiring new technologies, and deepening our understanding of the universe. The potential is infinite, much like the fractals themselves.

To wrap up, let's recall what Benoît Mandelbrot, the father of fractal geometry, once said: "Fractals are easy to explain, it's like explaining colors to someone. They're easy to explain, but they're impossible to define." Perhaps the real future of fractals lies in the journey—in the infinite exploration and the endless fascination they inspire.

As we continue our journey, armed with curiosity and wonder, we dive into the vast ocean of patterns and complexities, ready to unravel more fractal mysteries that the universe has to offer. So here's to the journey ahead and the infinite beauty of fractals!

Appendix

algorithm: A step-by-step procedure for calculations.

Cantor set: A fractal and an early example of a set that is "self-similar."

cardinality: A measure of the "number of elements" of a set.

chaos theory: A branch of mathematics focusing on the behavior of dynamical systems that are highly sensitive to initial conditions.

complex numbers: Numbers that consist of a real part and an imaginary part.

cosmology: The study of the origins and eventual fate of the universe.

dimension: An independent extent or quantity in a coordinate system or space.

Fibonacci sequence: A series of numbers in which each number is the sum of the two preceding numbers, usually starting with zero and one.

fractal: A shape that is self-similar, meaning it looks the same at any scale.

fractal dimension: A ratio providing a statistical index of complexity comparing how detail in a pattern changes with the scale at which it is measured.

fractal geometry: The study of mathematical shapes that are fractals.

golden ratio: An irrational mathematical constant, approximately 1.6180339887, which is often found in nature and is believed to create aesthetically pleasing compositions.

iteration: The act of repeating a process, often with the aim of approaching a desired goal or target.

Julia set: A set of complex numbers that, when applied repeatedly to a function, do not tend toward infinity.

Mandelbrot set: A set of complex numbers named after Benoît Mandelbrot, a mathematician who studied and popularized it.

number theory: The branch of pure mathematics devoted

to the study of the integers.

phyllotaxis: A term coined by Charles Bonnet (1720–1793), a Swiss naturalist and philosopher, to describe the arrangement of leaves on a plant stem.

punctuated equilibrium: A theory in evolutionary biology proposing that once a species appears in the fossil record, the population will become stable, showing little evolutionary change for most of its geological history.

quantum physics: A branch of physics dealing with phenomena on a very small scale, such as the behavior of subatomic particles.

self-similarity: A property where a shape is made up of smaller copies of the same shape.

Sierpiński triangle: A fractal and attractive fixed set named after the mathematician Wacław Sierpiński, who described it in 1915.

zoom invariance: A property of fractals where the same patterns recur regardless of how much you zoom in or out.

Appendix: People

Benoît Mandelbrot (1924–2010): A Polish-French-American mathematician and polymath with broad interests in the practical sciences. Known for developing a theory of roughness and self-similarity in nature and introducing the term "fractal."

Georg Cantor (1845–1918): A German mathematician best known as the inventor of set theory, which has become a fundamental theory in mathematics. Cantor established the importance of one-to-one correspondence between sets, defined infinite and well-ordered sets, and proved that the real numbers are more numerous than the natural numbers.

Wacław Sierpiński (1882–1969): A Polish mathematician known for outstanding contributions to set theory, number theory, theory of functions, and topology. The Sierpiński triangle, a fractal structure, is named after him.

Pierre Fatou (1878–1929) and **Gaston Julia (1893–1978)**: French mathematicians who independently described what are now known as Julia sets.

Edward Norton Lorenz (1917–2008): An American mathematician and meteorologist, and one of the pioneers of chaos theory. He introduced the strange attractor notion and coined the term "butterfly effect."

Leonardo Fibonacci (1170–1250): An Italian mathematician from the Republic of Pisa, considered to be "the most talented Western mathematician of the Middle Ages." The Fibonacci sequence is named after him.

Stephen Jay Gould (1941–2002) and **Niles Eldredge (born 1943)**: American paleontologists and evolutionary biologists who proposed the theory of punctuated equilibrium.

Euclid of Alexandria (c. 300 BCE): A Greek mathematician, often referred to as the "founder of geometry." His *Elements* is one of the most influential works in the history of mathematics.

Pythagoras of Samos (c. 570–c. 495 BCE): An ancient Ionian Greek philosopher and the eponymous founder of

Pythagoreanism. His theorem is widely used in mathematics today.

Albert Einstein (1879–1955): A German-born theoretical physicist, widely acknowledged to be one of the greatest physicists of all time. Einstein is best known for developing the theory of relativity.

James Webb (1906–1992): An American government official who served as Undersecretary of State from 1949 to 1950. He was the second Administrator of NASA from February 14, 1961, to October 7, 1968. Webb led NASA from the beginning of the Kennedy administration through the end of the Johnson administration, thus overseeing each of the critical first crewed missions throughout the Mercury and Gemini programs until days before the launch of the first Apollo mission. He also dealt with the Apollo 1 fire.

Timeline

Ancient Greece (~600 BCE): Pythagoras begins to develop Pythagoreanism, a philosophical and religious movement that greatly influences mathematics and abstract thinking. The Greeks lay much of the groundwork for geometry, a field of study that will eventually give rise to fractals.

300 BCE: Euclid, a Greek mathematician, writes the *Elements*, presenting the foundations of geometry.

1202 CE: Leonardo Fibonacci introduces the Fibonacci sequence in his book *Liber Abaci*.

1872: Karl Weierstrass presents an example of a function with no derivative, a concept that challenges the established understanding of the differentiability of mathematical functions.

1883: Georg Cantor introduces the Cantor set, an early instance of a fractal-like structure.

1898: Gaston Julia is born. He will later develop the groundwork for what will become known as Julia sets.

1906: Helge von Koch describes the Koch curve, a continuous curve notable for its self-similarity.

1918: Gaston Julia and Pierre Fatou introduce Julia sets after studying the iteration of rational functions, although without the aid of computers, they can't visualize them as we do today.

1924: Benoît Mandelbrot is born. His work will greatly contribute to the development of fractal geometry.

1952: Niles Eldredge is born. He and Stephen Jay Gould will later develop the theory of punctuated equilibrium in evolutionary biology, an area where fractal geometry is used.

1961: Edward Lorenz develops the foundations for chaos theory while studying weather patterns, uncovering strange attractors and the butterfly effect.

1975: Benoît Mandelbrot publishes *Les objets fractals: Forme, hasard et dimension*, a landmark work that popularizes fractal geometry. The term "fractal" is coined.

1979: Mandelbrot discovers the Mandelbrot set, an iconic fractal structure.

1982: The first conference on fractal geometry is held in New Haven, Connecticut, demonstrating the growing interest in the field.

1990s–2000s: Fractal geometry starts to find diverse applications in various fields, such as computer graphics, economics, physics, and medicine.

2010: Benoît Mandelbrot dies, marking the end of an era for fractal research.

2012–2023: Rapid developments and applications occur in the field of fractals, particularly in quantum physics, cosmology, AI and machine learning, and more. The James Webb Space Telescope is launched in 2021, promising to yield more discoveries that may impact fractal cosmology.

Bibliography

Cantor, Georg. "Über eine Eigenschaft des Inbegriffes aller reellen algebraischen Zahlen." *Crelle's Journal* (1874)

Crichton, Michael. *Jurassic Park*. Ballantine Books, 1990.

Eldredge, Niles., and Stephen J. Gould. "Punctuated equilibria: an alternative to phyletic gradualism," In *Models in Paleobiology*, edited by Thomas J. M. Schopf Freeman Cooper, 1972.

Euclid. (300 BCE). *Elements*.

Falconer, Kenneth. *Fractals: A Very Short Introduction*. Oxford University Press, 2013.

Feynman, Richard P. *QED: The Strange Theory of Light and Matter*. Princeton University Press, 1985.

Gleick, James. *Chaos: Making a New Science*. Penguin Books, 1987.

Hofstadter, Douglas R. *Gödel, Escher, Bach: An Eternal Golden Braid*. Basic Books, 1979.

Julia, G., and P. Fatou, (1918). "On the Iteration of Rational Functions." *Journal of Mathematical Studies* (1918)

Leibniz, Gottfried W. *Discourse on Metaphysics*, 1687

Lorenz, Edward N. "Deterministic Nonperiodic Flow." *Journal of the Atmospheric Sciences* (1963)

Lorenz, Edward N. *The Essence of Chaos*. University of Washington Press, 1993.

Mandelbrot, Benoît.B. *The Fractal Geometry of Nature*. W.H. Freeman and Company, 1983.

May, Robert M. "Simple mathematical models with very complicated dynamics." *Nature* (1976)

Peitgen, Heinz-Otto., Hartmut Jürgens, and Dietmar Saupe. *Chaos and Fractals: New Frontiers of Science*. Springer, 2004.

Penrose, Roger. *The Road to Reality: A Complete Guide to the Laws of the Universe*. Vintage, 2004.

Poincaré, Henri. "Sur le problème des trois corps et les équations de la dynamique." *Acta Mathematica* (1890)

Stewart, Ian. *Flatterland*. Perseus Publishing, 2001.

Turing, Alan. "The Chemical Basis of Morphogenesis." *Philosophical Transactions of the Royal Society B.* (1952)

Made in United States
North Haven, CT
21 October 2023